ANNALS OF
THE NEW YORK ACADEMY
OF SCIENCES

Volume 352

EDITORIAL STAFF

Executive Editor
BILL BOLAND

Managing Editor
JOYCE HITCHCOCK

The New York Academy of Sciences
2 East 63rd Street
New York, New York 10021

LONG-SPAN BRIDGES
O. H. Ammann Centennial Conference

OTHMAR HERMANN AMMANN
1879–1965

ANNALS OF THE NEW YORK ACADEMY OF SCIENCES

Volume 352

LONG-SPAN BRIDGES
O. H. Ammann Centennial Conference

Edited by Edward Cohen and Blair Birdsall

The New York Academy of Sciences
New York, New York
1980

ANNALS OF THE NEW YORK ACADEMY OF SCIENCES

VOLUME 352

December 31, 1980

LONG-SPAN BRIDGES *
O. H. Ammann Centennial Conference

Editors and Conference Chairmen
EDWARD COHEN AND BLAIR BIRDSALL

———◆———

* This series of papers is the result of a conference entitled Long-Span Bridges, O. H. Ammann Centennial Conference, held November 13 and 14, 1979 in New York City and sponsored by The New York Academy of Sciences.

Financial assistance was received from:

- AMERICAN BRIDGE, A DIVISION OF UNITED STATES STEEL
  CORPORATION
- AMERICAN INSTITUTE OF STEEL CONSTRUCTION, INC.
- AMERICAN IRON AND STEEL INSTITUTE
- BROWN BOVERI CORPORATION
- STEINMAN, BOYNTON, GRONQUIST & BIRDSALL, INC.
- WISS, JANNEY, ELSTNER AND ASSOCIATES, INC.

INTRODUCTION

Othmar Hermann Ammann is among the immortals of bridge engineering. The great and beautiful structures that he has bequeathed to us will serve and inspire many generations. He was a great innovator, and many of the techniques we take for granted in bridge design and construction today were originally conceived by him and incorporated into his structures.

It is most appropriate that outstanding engineers from around the world are assembled here in New York City to commemorate the 100th anniversary of O. H. Ammann's birth, by reviewing the history, current status, and future of long-span bridges. He devoted almost his entire career to the planning, design, and construction of long-span bridges, including three record-breaking spans.

Othmar Hermann Ammann was born in Schaffhausen, Switzerland on March 29, 1879 and received the degree of Civil Engineer from the Swiss Federal Polytechnic Institute in 1902. He came to New York in 1904 ". . . with only an engineering diploma in my pocket and little practical experience. I came here eager to learn. . . ." Thus began a career of achievement that ended only with his death in 1965, shortly after the dedication of the Verrazano Bridge.

When in 1924 he was appointed the first Chief Bridge Engineer of the newly created Port Authority of New York and New Jersey, he said, with great modesty and humility, "I felt myself saddled for the first time with heavy responsibilities, the discharging of which rested largely upon my own knowledge and judgment. However, I was fortunate in that, very wisely and with broad foresight, the Port Authority surrounded me with an exceptional group of consultants of high reputation and wide experience."

In 1931, within two weeks of each other, the most magnificent bridges of their type yet constructed, the George Washington Suspension Bridge and the Bayonne Arch Bridge, were opened to traffic.

From 1934-1939, Ammann served also as Chief Engineer of the Triborough Bridge and Tunnel Authority and directed the planning and construction of the Triborough and the Bronx-Whitestone suspension bridges. During this period he also played an important role in the design and construction of the Golden Gate Bridge in San Francisco.

In 1939, after retirement from both the Port Authority of New York and New Jersey and the Triborough Bridge and Tunnel Authority, O. H. Ammann went into private practice, investigating the Tacoma Bridge failure, conducting a study of the Brooklyn Bridge, and consulting in the design of the Delaware Memorial Bridge and the Mackinac Straits Bridge.

In 1946, at the age of 67, he entered into partnership with Charles S. Whitney to form the firm of Ammann & Whitney in the general practice of engineering, which included bridges, highways, airports, hangars, arenas, and special structures of all types throughout the world.

The firm's projects included the Throgs Neck Suspension Bridge in New York City and the Walt Whitman Suspension Bridge in Philadelphia.

The last and greatest project directed by O. H. Ammann, the Verrazano Bridge across New York Harbor with a main span of 4,260 feet, with two levels and 12 lanes of traffic, was opened on November 21, 1964, 6 months ahead of schedule and 10 months before his death.

O. H. Ammann said of the Verrazano-Narrows Bridge: "The planning and construction of the Narrows Bridge offered a great challenge to its planners and builders. Their satisfaction will be derived from its successful completion and from the great benefits it will bring to the community and the traveling public. Its planning and construction will be a testimonial to the collaboration of many public and private individuals and organizations, all of whom must share in the credit for its successful completion. It also reflects the great advances made in science and technology, without which its construction would not have been possible."

In addition to steel and concrete, O. H. Ammann has left guidance for engineers. ". . . more than ever our profession will be confronted with tasks which far transcend the accomplishments up to this time. They will call for further intensive study, research and the maintenance of highest standards by the profession. The road to success and leadership will be open to all who will apply effort, courage, and perseverance."

As we review the career of Othmar H. Ammann, we find an engineer of great vision, pursuing with single-minded devotion the concept of great and beautiful structures. His career spanned half a century, but its benefits will accrue to many generations.

———————◆———————

I wish to acknowledge the extensive assistance of Blair Birdsall throughout the conference.

Thanks are due to Anton Tedesko, Anthony R. Ameruso, Stanley Gordon, and Charles Scheffey for their assistance in chairing the conference sessions, and special appreciation goes to Professor H. H. Hauri for his excellent presentation on the history of bridge building.

I am grateful to Paul Duggan, Marie Mitchell, and Ann Balzano for their administrative and secretarial assistance during the conference and the production of this book.

Edward Cohen

New York, New York

LONG-SPAN SUSPENSION BRIDGES:
A BRITISH APPROACH

WILLIAM C. BROWN

INTRODUCTION

Many of the early river crossings for both road and railway constructed in Britain during the first half of the last century were of the classic suspension type. The growth of the railways system, and the consequent increase in unit loads transported by rail, quickly demonstrated that suspension bridges presented problems due to their relatively high flexibility and unacceptable distortion with the passage of freight trains.

The great railway engineers of the time turned their attention to more rigid bridge structures of modest span and to underwater bored tunnels for wider crossings. Thus we see the use of the hybrid arch suspension truss of 455-ft. span across the River Tamar at Saltash in S. W. England, designed by I. K. Brunel and completed in 1856, and Stephenson's Britannia Bridge with the deep box girders of 460-ft span over the Menai Straits between N. Wales and the Island of Anglesey completed in 1850 to link the Irish ferry terminal to the rail route to London.

Although the rail crossing of the Firth of Forth was originally proposed as a double-span suspension bridge by Thomas Bouch, it is doubtful if it would have been successful, and the majestic cantilever bridge with spans of 1710 ft designed by Sir Benjamin Baker and Sir John Fowler demonstrated that large steel truss bridges offered a very satisfactory and economic solution to the needs of the railways for most if not all the water crossings in Britain. It was completed in 1890.

Although the standard truss and plate girder spans which followed were necessary to fulfill the need for rigid rail expansion and were rather pedestrian in character, it is interesting to note that some of the principles of construction used in modern bridges can be seen in the Saltash, Britannia, and Firth of Forth rail bridges as well as other contemporary works.

The massive tubes of the Saltash bridge, for example, were elliptical in cross section, 16 ft 9 in broad by 12 ft 3 in deep, and stiffened in part. The box girder form and the longitudinal stiffening of the Britannia bridge were "rediscovered" a century later. The largest compression chords of the Forth Railway bridge were built up of curved plates to form circular tubes, up to 12 ft in diameter, while the smaller compression members were square or rectangular in section, stiffened longitudinally and transversely, much as in the modern steel box girder.

The development of the mass-produced motor car, and the subsequent slow decline of rail transport in the period before the last war, influenced bridge construction differently in North America and Europe. The need to effect road connections across wide waterways was a different proposition from the rail

---

William C. Brown, OBE, RDI, is a partner of Freeman Fox & Partners, 25 Victoria Street (South Block), London SW1H 0EX England.

0077–8923/80/0352–0001 $01.75/1 © 1980, NYAS

PLATE 1. The Royal Albert Bridge, Saltash, which links Devon and Cornwall.

PLATE 2. Britannia Bridge.

crossings and this, combined with a more adventursome spirit in the United States, meant that the progress towards successful long-span road bridges was left to American engineers to pursue. We all know how well they have succeeded in their task. Those who follow should never forget the debt owed to those pioneering engineers like O. H. Ammann: men who steadfastly challenged nature in their need to evolve practical solutions to satisfy the needs of society and who retained clear objectives, rejecting criticism on the way.

The construction of the George Washington Bridge in 1961, with its span of 3,500 ft, was such an advance over previous limits that it must always be regarded as a milestone in suspension bridge development which, in the future, will be remembered by those of us wanting to make comparable leaps in span.

UNITED KINGDOM CONCEPTS

It was that not until the late 1940s that we in Britain got around to thinking of road bridges with spans comparable to, if somewhat less than, that of the George Washington.

These were across the Severn Estuary separating England and Wales and located just north of Bristol and over the Firth of Forth near Edinburgh, close to the Forth Railway Bridge. Originally the central spans of both these two new bridges were to be the same, 3,240 ft. The final structures differ by 60 ft (19 m) and development of the basic design, particularly that of the suspended structure and cables, was considered to be applicable to both locations. The substructures, anchorages and towers were, for a variety of reasons, to be different.

A start on design was delayed, however, and it was not until 1956 that the definitive design of the first, that across the Firth of Forth, was initiated. In essence, it followed the general principles of an outline design prepared earlier but incorporating many developments in detail design and fabrication techniques so as to be compatible with the most advanced contemporary industrial processes.

The work of American engineers in this field had considerable influence on British engineers at the time and the experiences of the first Tacoma Narrows bridge considerably influenced the thinking of the late 1940s.

FORTH ROAD BRIDGE

Thus the depth of Forth Road Bridge stiffening truss was made one-hundredth of the main span, following the concept that such a ratio was fundamental to the achievement of aerodynamic stability.

FIGURE 1 shows the deck cross sections. The design approach is clearly shown.

The stiffening trusses and their associated bracing form is a structure that is separated from the decking system that might be said to follow the classic American approach. Nevertheless, this arrangement was evolved as a result of extensive model testing in wind tunnels during the early stages and only confirmed during the definitive design period.

The desire to test and check significant design changes, using models in wind tunnels, was an important factor influencing fundamental design changes at that time, though access to suitable tunnels was both difficult and limited.

However, by working in close contact with those who had been responsible

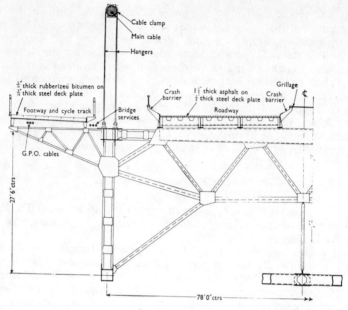

CROSS GIRDER OF MAIN SPAN

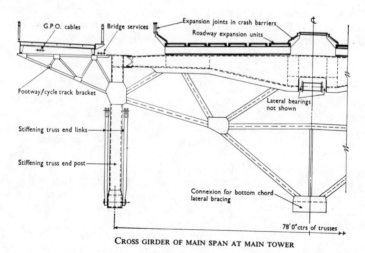

CROSS GIRDER OF MAIN SPAN AT MAIN TOWER

FIGURE 1.

for the studies on the first and second Tacoma Narrows bridges, aerodynamic research was carried out at the National Physical Laboratory in England. As a result of this work the deck of the Forth Road bridge was made torsionally stiff so as to raise the flutter speed and incorporated open spaces between the carriageways, footways and cycle tracks, and the associated structures primarily to deal with the motions induced by eddy shedding which were otherwise liable to be present.

FIGURE 2 shows how the chosen structure was evolved from several alternative, but unsatisfactory, solutions.

Following the requirements set by the earliest conception of the bridge, separate cycle and footway tracks were provided while the carriageways were limited to two lanes in each direction. These have proved quite adequate for the volume of motor traffic, which now averages about 25,000vpd.

By the mid-1950s methods of steelwork fabrication, which had not changed very much since the turn of the century, were undergoing change and the

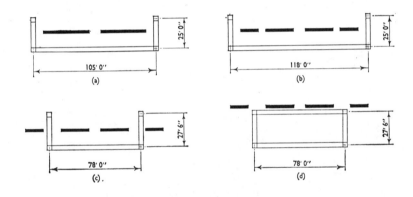

FIGURE 2. Evolution of stiffening truss and deck configuration.

change from riveted to welded fabrication moving at a pace. Indeed though the first tender designs were based on 20% of the work being riveted, details were soon changed to suit welding and only the truss brackets supporting the footway and cycle tracks remained riveted. All site connections were made with high-strength, friction-grip bolts except for a portion of the main span roadway decking, which was site welded.

The truss chords, formed using automatic machines from four plates welded at the corners, were made of high tensile quality steel and represented one of the earlier instances where such steel was welded. Thicknesses were kept to less than 1 inch wherever possible to help the steelmakers maintain the required mechanical properties yet keep the carbon below the 0.2310.25% range. (The development of more weldable steels was still a few years away, and in fact they were not available before the Severn bridge was under way.)

The problem of twist within the length of the chords, quite common at that time on welded box members, was accommodated at each bolted chord splice by drilling the holes using carefully prepared jigs.

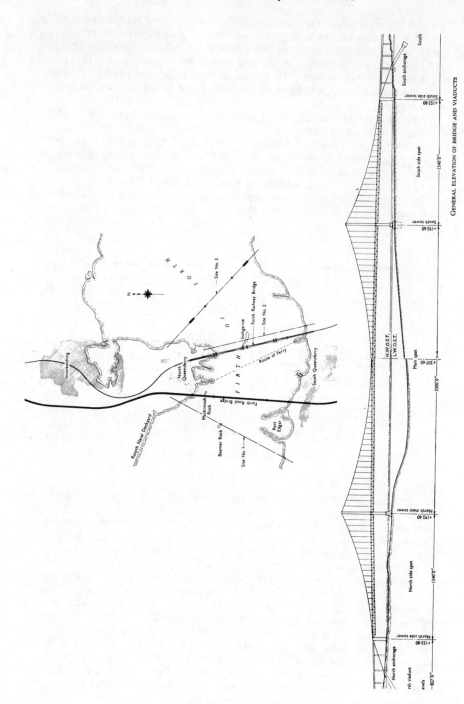

GENERAL ELEVATION OF BRIDGE AND VIADUCTS

PLATE 3. Forth Road Bridge.

TOWER DESIGN AND FABRICATION (FIGURE 3)

The towers represented quite a radical departure from previous designs and construction methods and, in part, reflected developments that had already taken place in the construction of very large columns in industrial buildings such as major power stations. Clearly, to follow such developments, the towers had to be plated and welded, but the number of plates and cells had to be kept small if the cost was not to be excessive. Considerable attention was given to the vertical lines of the tower as this was a significant design parameter. Previous experience has shown that good vertical alignment was possible if the ends of the welded tower cells were carefully milled after shop welding. This approach opened up the possibility of dispensing with splice covers by allowing the sections to bear directly at their ends. The problem of twist in the box sections was also overcome if they were provided with end rings.

With such a method of connection between the boxes, the remaining problem was the uplift experienced under wind pressure before sufficient weight of the cable and suspended structure was in place. The incorporation of long very high tensile screwed rods passing through bolt brackets, and pretensioned with nuts at their ends, effectively solved this problem. The individual boxes at each level were linked with vertical plates attached by bolting (FIGURE 3).

The first tender design envisages six boxes, the size of each limited to about 6 ft 6 in by 4 ft 6 in, with the whole cross section containing 11 cells. In the event, however, with the availability of larger horizontal milling machines, the number of welded boxes was reduced to three and the number of cells to five.

Since the essential purpose of the tower is to carry loads from the cable to the pier, the most efficient solution is one where the overall axial stress is maintained at as high a level as is reasonable (FIGURE 4). In practice this means using high quality structural steel and keeping the bending stresses down to a minimum. For practical convenience during erection, the tower bases are fixed to the piers, but any disadvantage is more than offset by restricting the width of the tower in the longitudinal direction of the bridge to that needed for ultimate stability and reasonable erection stiffness. Not only is it then practical to fix the saddles to the tower top, but a considerable overall economy results. FIGURE 5 demonstrates the effect of bending due to deflection on the available axial stress.

CABLES

All the major British designed suspension bridges have main cables that are of the built-in-situ parallel wire type pioneered in the USA by J. A. Roebling for the Brooklyn Bridge. The contractor responsible for the construction of the Forth Road Bridge sought the advice and experience of the Roebling Company and were indebted to Mr. Blair Birdsall and his team, particularly Mr. A. W. Hills, for their necessary and constant guidance. Subsequently, British contractors have been able to act on their own, but the system used has always been essentially "Roebling," with the strands set in a "vertical" rather than a "horizontal" hexagonal form.

The cable bands generally follow American practice, but the clamping bolts are different, being of waisted shank screwed rods of very high tensile strength and working to a much higher stress than hitherto (FIGURE 6).

NORTH TOWER WITH 9 OUT OF 11 SECTIONS ERECTED, SHOWING CLIMBING
STRUCTURE AND MAN/MATERIAL HOIST

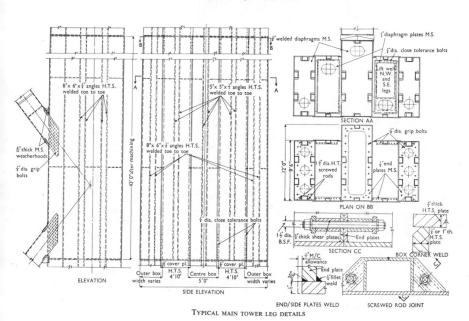

TYPICAL MAIN TOWER LEG DETAILS

FIGURE 3.

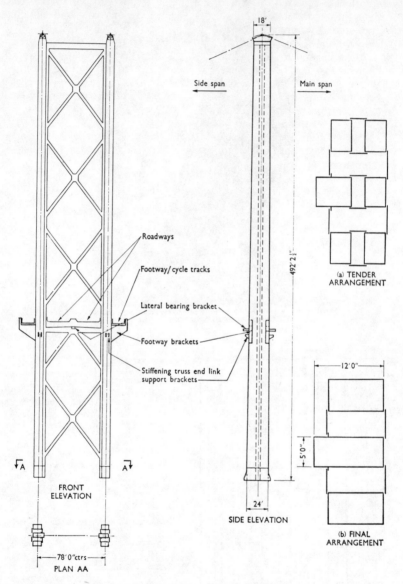

Roadways

Footway/cycle tracks

Lateral bearing bracket

Footway brackets

Stiffening truss end link
support brackets

Side span

Main span

18'

492' 2 ½"

(a) TENDER
ARRANGEMENT

12' 0"

5' 0"

(b) FINAL
ARRANGEMENT

FRONT
ELEVATION

SIDE ELEVATION

24'

A

A

78' 0"ctrs

PLAN AA

COMPARISON OF TENDER AND FINAL TOWER LEG CELL ARRANGEMENT AND
GENERAL ARRANGEMENT OF A MAIN TOWER

FIGURE 4.

The advantage claimed for is not only a reduction in the number of bolts and clamp length required, but there is a smaller reduction in clamping force for a given finite contraction of the cable likely to occur with load and time. Experience seems to support this view.

The design and form of any anchorage depends greatly on the local ground conditions and general topography, so it is the method of linking the cable to the anchorage that can be considered in isolation. At the time, we were well aware of the American use of eye-bar configurations and the means of adjustment. We were equally aware that forging such bars was outside the easy capacity of British construction companies.

An alternative solution had to be evolved. This is shown in FIGURE 7.

The interface slab is prestressed to the anchorage face and large diameter screwed rods are later passed through it and used to fix the double strand shoes. Strand adjustment is effected on the screwed threads.

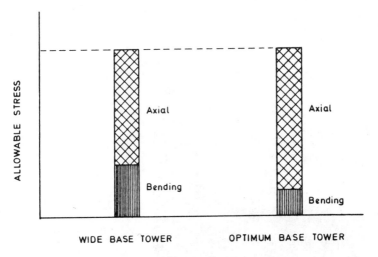

FIGURE 5.

This arrangement was originally developed for the Forth Road Bridge and had been adopted on all subsequent British designs.

SEVERN AND LATER BRIDGES

Perhaps the most interesting development was the shift away from the lattice truss form of stiffening girder to the enclosed shaped box section deck, first used for the crossing of the River Severn and later, in adapted form, at Little Belt in Denmark, across the Bosporus in Turkey, in Korea and for the Humber crossing currently under construction.

It had long been realized that, as the span increases, the unit dead weight suspended structure remains constant while the weight of the cable increases, thus constituting an increasing proportion of the total weight. This means that

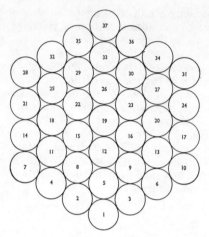

PATTERN OF STRANDS IN CABLE BEFORE COMPACTING

COMPACTING MAIN CABLE

FIGURE 6.

ADJUSTING ONE OF THE CABLE STRANDS BY JACKING THE STRAND SHOE WHICH
CONNECTS IT TO THE ANCHORAGE

FIGURE 7.

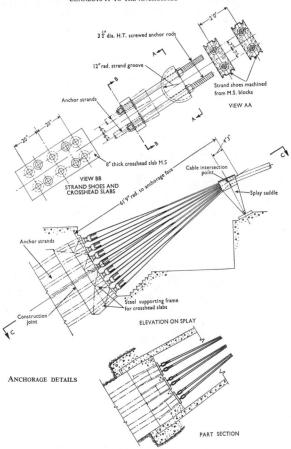

ANCHORAGE DETAILS

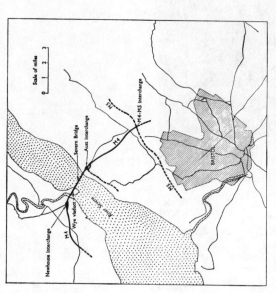

Map showing bridge location

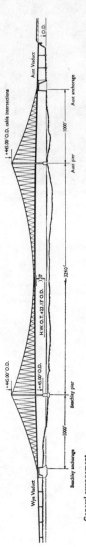

General arrangement

Plate 4. Severn Bridge.

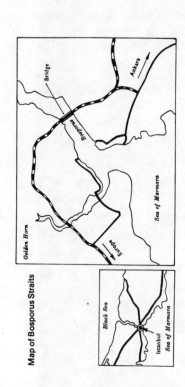

Map of Bosporus Straits

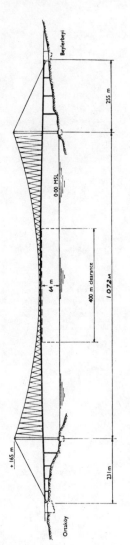

General arrangement of bridge

PLATE 5. Bosporus Bridge.

the dead load tension within the cable is higher and the bridge stiffness so derived from this tension more effective. So much so in fact, that, for spans over 2000 ft (600 m), the presence of the "stiffening" girder was of little benefit in controlling even load distribution of loads. There is of course an apparent balance between increasing the sag and reducing cable size and stiffness, which has to be compensated for by adding rigidity in the decking, and the opposite approach, i.e. providing minimum deck stiffness consistent with the proper distribution of traffic loads to the cables and control of the cable sag to provide cable tension-stiffness. The George Washington bridge as originally constructed effectively demonstrated the latter principle.

The deeper truss structures provided a convenient means of incorporating torsional rigidity into the system, a feature considered necessary to control aerodynamic stability, and if sufficient aerodynamic stability was maintained, any effective reduction due to a shallower girder needed to be compensated for in other directions.

The integration in the Severn bridge design of the roadway decking into the effective action was an obvious benefit, while the incorporation of the chords, diagonal and lateral bracing into the minimum section solid webs, and bottom flanges of the plated box did in fact reduce the amount of steel while maintaining a torsional strength comparable with previous designs.

The vital question was the possible aerodynamic behavior of a shallow yet torsionally stiff suspended deck section.

The sections finally adapted for all three British-designed bridges with box deck girders are shown in FIGURE 8. The evolution of end shapes for these sections was a result of wind tunnel tests. FIGURE 9 shows some of the shapes examined.

At the time that the Severn was designed, late 1950s-early 1960s, we did not appreciate the full significance of the degree of torsional stiffness required as we think we do nowadays. More recent development work has shown that the effect of shape and torsional rigidity on the flutter speed for given bridge parameters is similar to that shown in FIGURE 10.

The influence of section shape as well as of stiffness is apparent. It should also be noted that there is a separation of flutter speed frequencies due to the disposition of mass about the center of angular rotation. Separation is increased by spacing the cables further apart but reduced by adding mass, such as footway and cycle tracks, outside the line of the cable. Thus it would appear that the Severn-type cross section is not an optimum in this report, but it seems that any difficiency is partly compensated by improved structural and aerodynamic efficiency.

The wind tunnel tests, while demonstrating the stability for flutter, did indicate a possible minor instability caused by eddy shedding at a narrow band of wind speeds around 78mph. These oscillations could be damped quite easily, but since it was the intention to save weight and cost by welding the whole of the suspended structure, little structural damping could be expected. Some external damping was needed and this was obtained by inclining the hangers and utilizing frictional effects within the strand to dissipate energy. The permitted inclination was limited to that where the hangers were unlikely to lose tension during the passage of traffic. This limit was set at 65°.

After various trials, the best form was deemed to be a single round strand of 180 wires layed up in a short lay (i.e. steeper spiral).

Being a strand, albeit more flexible than normal, clearly the hangers could

Forth Road Bridge

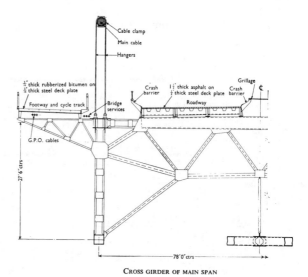

CROSS GIRDER OF MAIN SPAN

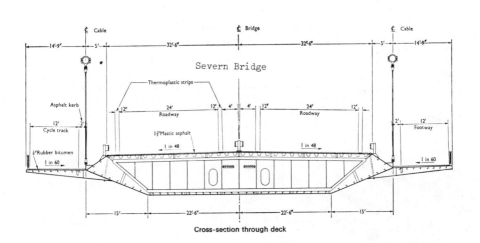

Severn Bridge

Cross-section through deck

Bosporus Bridge

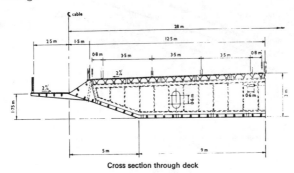

Cross section through deck

FIGURE 8.

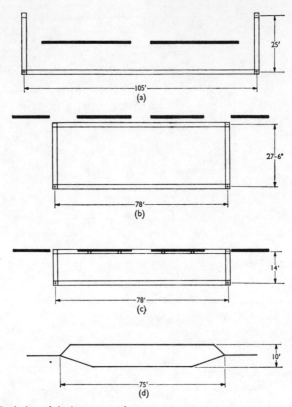

Evolution of deck cross section

FIGURE 9.

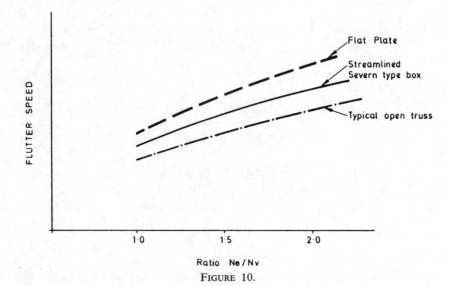

FIGURE 10.

not be hung around the cable out would need to be socketted at both ends and connected to the lower half of the cable band, which was now set with the clamping bolts vertical. Although requiring a few more bolts than otherwise, there were advantages in that the castings were easier to make and fewer variations were needed. The deck lifting gantries could also run more easily along the top of the cables. This arrangement of cable bands is shown in FIGURE 11.

One of the objectives in developing the box deck sections was to devise a design which was not only lighter, and therefore more economical, but to prepare a design in keeping with the latest industrial practice. Truss and lattice construction is suited essentially to bolted or riveted fabrication and, as explained earlier, this type of work was rapidly becoming uneconomic in Britain.

On the other hand, welded stiffened plate work was simple and cheaper, once the technique had been acquired. It also lent itself to more repetition and automation.

The multi-cell towers of the Forth Road Bridge had proved satisfactory but could be simplified. For the Forth bridge towers design had been based on large box section columns used in power stations and, on reflection, it seemed logical to replace the multi-cell arrangement by one large cell if such could be reduced to four transportable units.

The stiffening of wide plates no longer presented the problem it might have done a few years earlier, so the section shown in FIGURE 12 was devised.

One particularly interesting feature is the detail used to connect the four panels together. The vertical connecting plates on the Forth towers were bolted to the box units, conventionally with the bolt head outside and the nut inside. This required access all round the tower and a special access platform, giving protection against the frequent inclement weather, had to be provided. By using a re-entrant angle, however, the bolts are not only accessible for placement from the inside, but the feature provides an interesting line and appearance to the tower legs.

Similarly the deep portal web plates are connected internally and the re-entrant angle feature used to break up the otherwise deep flat surface, on which all the stiffener welding would have been noticeable. Other aspects of the design are shown in FIGURE 13.

The improved wider and slightly shallower deck of the Bosporus crossing, in spite of its increased span, provided somewhat better aerodynamic stability feature compared with the Severn. For the still longer span over the Humber (1410 m) with its narrower roadways, the depth was increased by 33% to provide adequate frequency separation, but this has increased the weight and cost.

Concrete is a material ideally suited to withstanding compressive forces and, in the absence of seismic conditions, its use for the Humber towers is appropriate. However, steel towers, shaped to reduce the amount of additional stiffening, and using better quality steels in thicker sizes, make current comparisons regarding the preference for steel or concrete less conclusive.

CURRENT THINKING

The reasoning behind the Severn concept is now 20 years old, and it would be surprising if improvements could not be made. There are many wide rivers

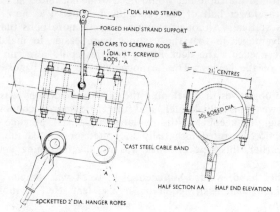

Cable band details

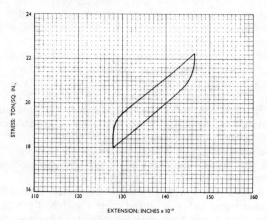

Typical hysteresis curve for hanger strand

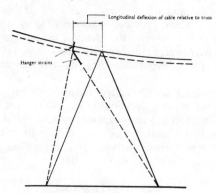

Hanger strain due to bridge oscillation

FIGURE 11.

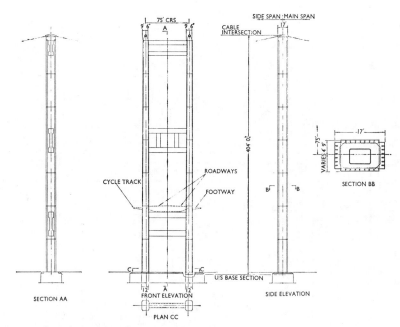

General arrangement of tower

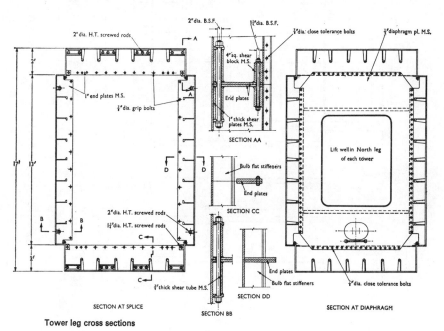

Tower leg cross sections

FIGURE 12.

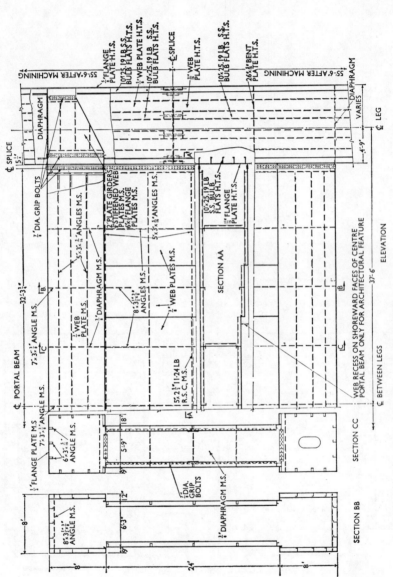

Typical tower leg and centre portal beam details

FIGURE 13.

and waterways waiting to be bridged, but such crossings are feasible only if the cost is right.

Thus not only must the material content be kept down but the labor cost and unit price as well. This would seem to point to greater repetition of parts, more automation, and inevitably a move away from field assembly to the fabricating shop. For longer spans, if an increasing amount of torsional stiffness is not to be required, the aerodynamic torsional-distributing force must be reduced at source.

A solution to this requirement does exist that seems to suit other needs as well. By combining the streamlining at Severn with the deck openings used

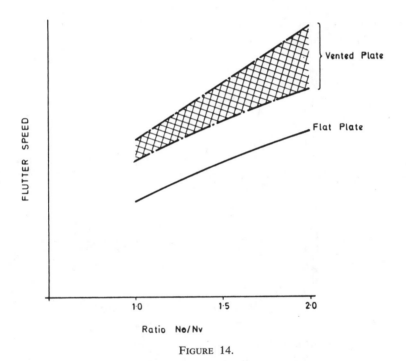

FIGURE 14.

since Tacoma, and carefully tuning the shape of the openings aerodynamically, it is possible to raise the critical wind speed above that of flat plate—see FIGURE 14. The benefit is largely destroyed if the undersides of the units are interrupted by conventional stringers etc., or if trusses are added, but these are really unnecessary. This concept, as well as being more efficient, has many constructional advantages and the proposed bridge of 3,300 m across the Messina Straits (FIGURE 15) shows how this can be applied.

Ammann came close to this solution in the original George Washington bridge. We now understand why it worked. I can apply it to larger works. I hope he would have approved.

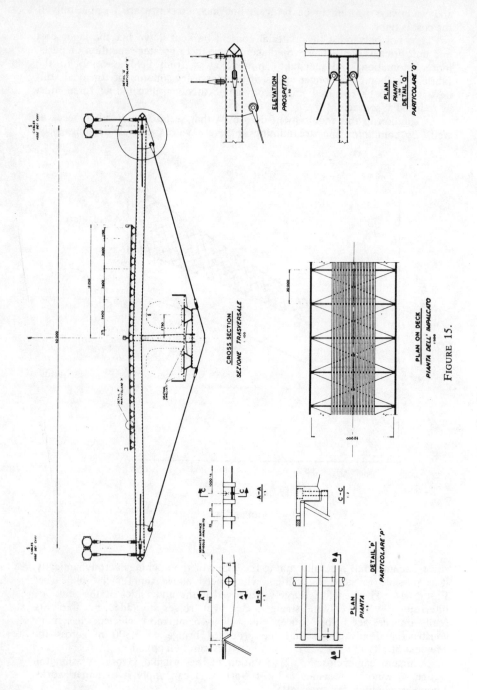

FIGURE 15.

LONG-SPAN SUSPENSION BRIDGES: THE AMERICAN APPROACH

EDWARD COHEN, ALLEN M. CUSTEN, AND FRANK L. STAHL

The center of gravity of long-span suspension bridge construction, which moved from Europe to the United States in the second half of the 19th century, has now, in the second half of the 20th century, returned to Europe and spread to the Far East. As the century of American predominance has brought numerous significant advances in design and construction techniques it is now quite appropriate to review the long-span suspension bridge experience in this country.

PLANNING

The planning of a long-span bridge project, or of any bridge project for that matter, has become infinitely more complicated than, say, in the days of the Brooklyn Bridge. While the obtaining of proper financing was then almost the only prerequisite for the design and construction of a bridge project, funding today is only one of many considerations that enter into the planning process.

Normally, bridge projects of this size can not be financed with public funds through the usual channels of state highway departments. Therefore, when a large bridge crossing is indicated, the first step is usually the creation by state or city legislature of a semi-autonomous authority that can finance the project with private capital through bond issues. Since sound economic policy demands that the project be self-supporting, the necessary revenue to repay the bonds is derived from tolls paid by the traffic using the bridge. One of the earliest efforts in the planning stage is, therefore, a traffic origin and destination survey which establishes the definite need for the project and from which present-day as well as future traffic usage of the bridge can be forecast. The decision on the number of lanes required to accommodate present as well as anticipated future traffic is a crucial one for the economic viability of the structure.

Federal law now requires a thorough study of the proposed project's impact on the environment, and no work can be started until it has been clearly established that the benefits of the project outweigh any detrimental effects on the natural surroundings or the economy of the region. While the legal ramifications of environmental consideration are relatively new, improvement of the environment in connection with the construction of large bridge projects has been a policy of responsible planners for a long time, as evidenced by the construction of parks and recreational facilities along the approach right-of-way of many bridges in New York City.

Selection of the proper alignment must take cognizance of economic considerations, particularly in urban areas. The relocation of residents and businesses along the approaches and the removal of properties from the tax roll

Edward Cohen and Allen M. Custen are partners and Frank L. Stahl is a senior associate in the consulting engineering firm Ammann & Whitney, 2 World Trade Center, New York, New York 10048.

0077–8923/80/0352–0027 $01.75/1 © 1980, NYAS

must be kept to a minimum; yet, the bridge approaches should be direct, rather than circuitous, to minimize construction cost and result in minimal vehicle operating costs.

Since most long-span bridges cross navigable waters, tower pier locations must clear channel and pierhead lines and permission must be obtained from the appropriate authorities both for the horizontal distance between tower piers and for the vertical clearance under the structure to avoid any interference with ship traffic. Once the center span of the bridge has thus been established, the sidespan length becomes a function of topographical features and of proper balance with the main span. Longitudinal grades must find an economic optimum between vehicle operation and lengths of approach structure to connect the bridge to surrounding grade.

The final determination of the bridge location and span arrangement can become a very complex task. It involves foundation explorations, very preliminary design studies of various alternate schemes, and order-of-magnitude cost estimates. For the Verrazano-Narrows Bridge, for instance, more than 20 combinations of tower and anchorage positions were studied before a satisfactory span arrangement was found, even after the final location of the bridge had been optimized to a very narrow band.

DESIGN

For the structural engineer, the design and construction of a long-span suspension bridge is probably the most fascinating possible bridge project. From an engineering point of view, progress in bridge building is broadly characterized by increases in the span length and traffic capacity, because engineering difficulties multiply progressively with these increasing proportions. Each structure that projects its span or capacity materially beyond precedents marks a new milestone in the progress of bridge construction.

In judging the merits of a bridge, consideration must be given to the era when it was built. Allowances must be made for the improvements in materials, machinery, and methods of construction. Materials available to today's engineer far exceed the strength and quality of those of 50 or 100 years ago and make our modern long-span bridges possible. The ultimate lengths of a suspension bridge main span will depend on the strength of available cable wire material and economic factors. Since for long spans a considerable portion of the available strength of the cable wire will be needed to carry itself, the remaining portion of available strength to carry the "payload"—the traffic-carrying suspended structure—will become smaller with ever increasing span lengths. It is generally accepted that span lengths in the 10–12,000-ft range present the ultimate economic limit of the current state of the art.

Design of the cables and stiffening trusses is still based on the deflection theory developed by Prof. Melan in Vienna [1] in the latter part of the 19th century and first applied in this country in the design of the Manhattan Bridge by Leon Moisseiff.[2] Extensive research work has led to refinements in theory and now permits a much more elaborate stress determination, particularly since the advent of electronic computers. Prior to that, the designer had to rely on time-consuming manual calculations with approximating assumptions or on elaborate model testing. With the help of this new electronic tool, bridges can be treated as three-dimensional structures with integral action and previously neglected secondary stresses can now be accurately determined. This results in a better utilization of materials and more economic construction.

For other parts of the bridge structure—roadway and supports, curbs, railings—design requirements for a bridge constructed in this country are prescribed in the Specification for Highway Bridges of the American Association of State Highway and Transportation Officials (AASHTO).[3] Allowable stresses for materials are also specified by AASHTO. For bridges not subject to federal regulations, design requirements and allowable stresses must be determined in accordance with local usage and the desired life of the structure.

Today, aerodynamic studies are an integral part of the design process. The collapse of the Tacoma Narrows Bridge in 1940 focused the concern of the design profession on the potential aerodynamic instability of long-span bridges and, for the first time, concentrated efforts were undertaken to gain a better understanding of this disturbing phenomenon. O. H. Ammann, who headed the team including Theodore Von Karman and Glenn Woodruff that reviewed the Tacoma disaster for the Federal Works Agency,[4] was instrumental in forming the Advisory Committee on the Investigation of Long-Span Suspension Bridges of the Public Roads Administration,[5] which included a host of well-known scientists and engineers of that era. The work of this group and the resultant recommendations must be considered a milestone in suspension bridge engineering.

In spite of the considerable amount of research performed in the last 30 years, a strictly scientific and accurate approach to the design of a long-span suspension bridge against dynamic wind action is still not available. The designer must rely in part on empirical information derived from the behavior of a variety of existing structures under long-time exposure to wind action and, primarily, on model testing. Improved wind tunnel techniques of producing and observing simulated wind conditions at the bridge site on a 3-dimensional model of the bridge still provide the most reliable estimate of the anticipated behavior of a proposed structure under adverse aerodynamic conditions.

If the proposed structure is located in a seismically active zone, additional consideration must be given to the forces acting on the bridge as a result of the strongest earthquake forecast for this particular location. Recent advances in earthquake engineering and research have resulted in the inclusion of minimum design parameters in the current AASHTO Specifications for Highway Bridges.[3] For long-span bridges, however, the design must be supplemented by studies of longitudinal ground motion excitations and out-of-phase excitation of towers and anchorages.

Last, but not least, consideration must be given to aesthetic factors in the design of the structure. Great bridges become historic landmarks, and their appearance influences their surroundings. While engineers are generally possessed with a strong sense of utility and are inclined to justify the appearance of any structure from the economic and scientific point of view, there is now a marked and welcome recognition of the need for proper aesthetic treatment of such large public projects. Even though the principal lines and proportions of the structure are primarily dictated by the fundamentals of strength and stability as well as by local geographical and topographical conditions, sufficient latitude remains for the architect or engineer to apply his own sense of beauty in determining the final aesthetic appearance of the structure. Consequently, some of our modern large bridges have been acclaimed by the public as much for their beauty as for their purely utilitarian and scientific function.

One of the first and major problems confronting the designer of a large bridge is to decide on the type of structure that would be economically and aesthetically most suitable for the particular location. Today it is well recognized that for very long spans the suspension bridge type, when properly designed, offers such outstanding merits that its superiority over any other type of structure is obvious. In determining the optimum cross section, traffic capacity, and economic construction the designer normally must make a decision among various alternative possible schemes. He must consider such diverse items as the proportions of the depth and width of structure in relation to the span; the currently required traffic capacity versus the economic benefits of providing for an estimated capacity forecast many years into the future; the materials and fabrication capabilities available within a convenient distance from the bridge location; and the probable ways of erecting the structure within the geographical and topographical constraints of the site.

A good example can be provided by the studies made for the Verrazano-Narrows Bridge (FIGURE 1). Many alternatives were investigated, among them arrangements for a single deck 6-lane capacity and for a 12-lane capacity on two decks. The final selected scheme with two 6-lane roadways for mixed traffic on two decks presents a radical departure from previous norms and may, in retrospect, have been one of the most significant and least publicized contributions of modern American suspension-bridge engineering. Prior to the Verrazano-Narrows Bridge, the customary framing of the longitudinal deck stringers served merely to carry the deck dead and live loads to the supporting members. Participation of the stringers in the overall action of the superstructure was carefully avoided by making them discontinuous with sliding connections at the deck expansion joints, spaced 40–50 feet apart.

In the Verrazano-Narrows Bridge design developed by O. H. Ammann every effort was made to integrate the longitudinal stringers with the deck structure (FIGURE 2). To accomplish this the stringers were placed in the planes of the upper and lower chords of the stiffening trusses and made continuous from tower to tower for a length of over 4,000 feet. The two stiffening trusses, the stringers of both decks, and the two lateral trusses in the plane of each deck are all rigidly connected to the rigid-frame floorbeams and thus form an integral part of a continuous laced-wall tubular structure in which the floorbeam frames act as heavy diaphragms. In this unique design all parts participate in resisting the vertical, torsional and lateral forces from moving loads and wind.

Stability of the suspended structure in high winds is important for the safety of the structure itself as well as for the safety of, and the continuous use by, the traveling public. Wind tunnel tests [6] and actual service experience have proven the excellent aerodynamic quality of the Verrazano-Narrows Bridge cross section. The openness of the structure permits a reasonable circulation of the wind stream and its effect on traffic is minimized. In contrast, while the winged-box section design is stable under severe wind conditions, the wind stream appears to be deflected and concentrated into the traffic paths, thus occasionally reducing the usefulness of the crossing in high winds.

The advantage of the Verrazano-Narrows deck section can best be illustrated by the following. When the bridge across the Bosporus was in the planning stage a proposal for a bridge using a cross section similar to the Verrazano Narrows design was submitted to the owners. A single 6-lane deck was proposed for initial construction, but the bridge would have had provisions for future

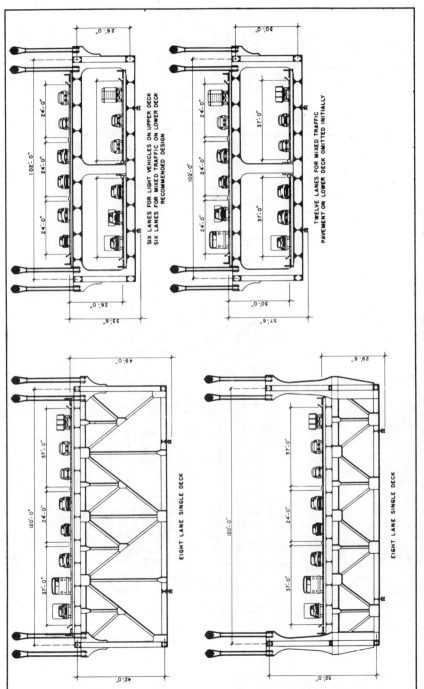

FIGURE 1. Verrazano-Narrows Bridge—alternate schemes.

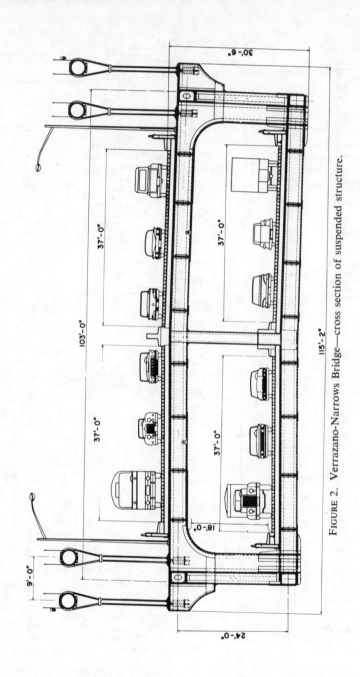

FIGURE 2. Verrazano-Narrows Bridge—cross section of suspended structure.

expansion with an additional 6-lane roadway on a lower deck. This proposal was rejected in favor of a 6-lane winged box design, which was eventually built. Within a few years of the opening of this bridge, traffic had increased to such proportions that a second parallel bridge is presently in the planning stage at an estimated cost of 90 million dollars. Accommodation of this traffic on an added lower deck of a Verrazano-Narrows type deck structure would have been possible at a fraction of this cost.

The Verrazano-Narrows deck structure is somewhat heavier than an equivalent capacity winged-box design (FIGURE 3). This increased weight is due to the use of an independent roadway—in this case a concrete-filled grid—as compared to the orthotropic floor construction of the winged-box section. Until now, orthotropic deck roadways have not been used in long-span suspension bridges in this country because of a preference for avoiding participation of

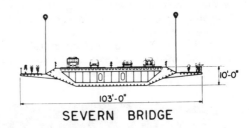

SEVERN BRIDGE

FIGURE 3. Severn Bridge and Verrazano-Narrows Bridge cross sections.

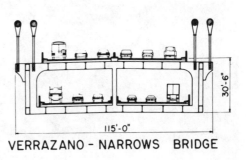

VERRAZANO - NARROWS BRIDGE

the roadway deck itself in the complex interaction of the cables and suspended floor system. In the orthotropic floor, the steel deck is continuous for the length of the span and acts as top flange of the steel construction. As such, it fully participates in the stress flow of the suspended system. The effect of these stresses on the roadway paving bonded to the steel plate greatly reduces the useful life of the paving. The inconvenience and cost of maintenance and repair of an orthotropic floor under today's crowded traffic conditions of American bridges fully justifies the preference for separating the roadway deck from the suspended floor systems, even at the cost of increased weight.

From a strictly technical point of view, orthotropic floor construction on a tubular truss-type bridge similar to the Verrazano-Narrows Bridge is entirely feasible. Its acceptance will depend on local economic and traffic conditions. Savings in weight resulting from orthotropic construction must not be offset by increased fabrication, shipping and erection costs. Repair or replacement of the roadway paving must be feasible without objectionable effects on traffic and at a reasonable life-cycle cost.

CONSTRUCTION

The great advances made in this century in the ever-increasing lengths and capacity of suspension bridges has been given impetus by the continuous improvement of available materials and innovations in construction methods.

Thanks to the continuous research performed by our major steel companies, we have today materials that provide greatly increased strength properties compared to those available in the days of the Brooklyn Bridge or the George Washington Bridge. This research has also given us a great variety of longer lasting materials for diverse applications, particularly suitable for welding, resistant to extreme temperatures, and capable of facilitating various fabrication processes. Similar advances have been made in the quality of cements and concrete admixtures.

In the shop-fabrication of structural steel members, important improvements have been made in the past 20–30 years that greatly facilitate the construction of large bridges (FIGURE 4). Computer-controlled machines that guarantee a

FIGURE 4. Verrazano-Narrows Bridge—tower fabrication.

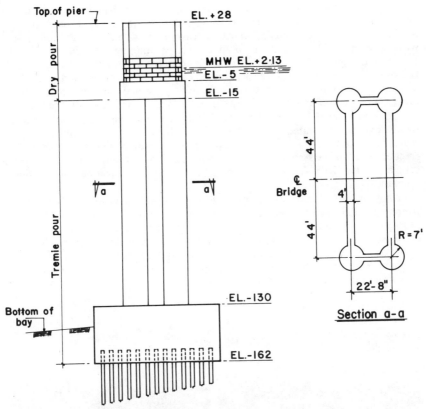

FIGURE 5. Newport Bridge—tower foundations.

remarkable degree of accuracy in the fabrication and assembly of individual steel members have been introduced. Powerful machines now permit the fabrication of individual members of ever greater size and weight and their transportation to the construction site.

When we compare the present-day erection of large bridges with that of 50 or 100 years ago, we notice striking improvements in the speed with which foundations are sunk to ever greater depths and enormous masses of steel and concrete are asembled and erected in the field. A few examples will illustrate these great advances.

Whereas in times past good and simple foundation conditions invariably were a major factor in locating a bridge crossing, today adverse foundation conditions merely pose a challenge which must be overcome. The tower foundations of the Newport Bridge across Rhode Island's Narragansett Bay are a case in point (FIGURE 5). A water depth of 160 ft, 5-knot tides, and treacherous and unpredictable weather conditions were overcome by unprecedented construction procedures which included underwater driving of piles, the floating-in and underwater setting within close tolerances of the prefabricated steel forms for pile cap and tower piers with the help of floating equipment of 500-ton lifting capacity, and the placement of 90,000 cubic yards of tremie concrete.[7]

The speed of cable erection also provides a good measure to demonstrate the increased capability of the construction industry. When first used by the Roeblings, wire was carried across the span from one side only, one loop or two wires at a time, at a speed of about 350 ft per minute. At the Verrazano-Narrows Bridge (FIGURE 6), wire was placed from both anchorages, simultaneously by four spinning wheels, carrying 2 loops or 4 wires each at a speed of 1200 ft per minute. This rate of cable spinning was double that of any prior bridge construction and about thirty times that of the Brooklyn Bridge.

On some recent bridges, the Newport Bridge and the Second Chesapeake Bay Bridge, prefabricated parallel wire strands have been used in the cable construction (FIGURE 7). This new method provides another tool of speeding field erection but appears to be limited—depending on sidespan proportions—to bridges in the 2000-ft mainspan range because of fabrication and shipping limitations.

Another recent innovation has been the development of a plastic cable wrapping. This system, as installed on the Second Chesapeake Bay Bridge (FIGURE 8), replaces the conventional soft-wire circumferential wrapping and consists basically of a neoprene elastomeric sheet protected by hypalon paint. Compared to the wire-wrap system, this new system provides a weight saving of about 75%, not insignificant in long-span bridges. Service experience of the new wrapping system is reported to be good.

FIGURE 6. Verrazano-Narrows Bridge—cable spinning.

FIGURE 7. Prefabricated parallel wire strand.

Finally, greatly improved capacity of erection equipment now permits erection of large preassembled sections of the suspended structure, contrasting sharply with the piece-by-piece member erection used as late as the 1950s. At the Verrazano-Narrows Bridge (FIGURE 9), 100-ft long sections of the entire floor structure, weighing close to 400 tons, were preassembled off site, floated in on barges and lifted into place as one unit.[8] The effect of such erection procedure on project completion and on savings in the financing of construction costs need hardly be emphasized.

CONCLUSION

The engineering profession and the construction industry can derive great pride and satisfaction from the tremendous progress made in suspension bridge

FIGURE 8. Chesapeake Bay Bridge—plastic cable wrapping.

FIGURE 9. Verrazano-Narrows Bridge—floor erection.

38

design and construction since the days of John Roebling. Intensive research guarantees that this progress will continue in the future.

REFERENCES

1. MELAN, J. 1888. Eiserne Bogenbrücken und Hangebrücken. Wilh. Engelmann. Leipzig, Germany.
2. MOISSEIFF, L. S. 1925. The towers, cables and stiffening trusses of the bridge over the Delaware River between Philadelphia and Camden. J. Franklin Inst. **200**, No. 4: 436–466.
3. AMERICAN ASSOCIATION OF STATE HIGHWAY AND TRANSPORTATION OFFICIALS. Standard Specification for Highway Bridges. Washington, D.C.
4. AMMANN, O. H., T. VAN KARMAN & G. B. WOODRUFF. The Failure of the Tacoma Narrows Bridge. A report to the Hon. John M. Carmody, Administrator, Federal Works Agency, Washington DC, by Board of Engineers (Othmar H. Ammann, Theodore Van Karman, Glenn B. Woodruff), dated March 28, 1941.
5. Private correspondence.
6. BRUMER, M., H. ROTHMAN, M. FIEGEN & B. FORSYTH. 1966. Verrazano-Narrows Bridge: Design of superstructure. J. Construction Division, Am. Soc. Civ. Eng. **92**, No. CO 2:57–66.
7. HEDEFINE, A. & L. G. SILANO. 1968. Newport Bridge foundations. Civil Engineering, Am. Soc. Civ. Eng. of October 1968, pp. 37–43.
8. KINNEY, J., H. ROTHMAN & F. STAHL. 1966. Verrazano-Narrows Bridge: Fabrication and construction of superstructure. J. Construction Division, Am. Soc. Civ. Eng. **92**, No. CO2.

MAIN CABLES AND APPURTENANCES

A. I. Zuckerman

Cables are the main carrying members of suspension bridges. Made of drawn steel wire, the steel is used in its strongest and most efficient manner.

Wire drawing is an ancient art. Used in jewelry, drawn gold and silver wire is found among the artifacts of the ancient Egyptians. Drawn steel wire was invented in 1706. The wire is drawn through progressively smaller and smaller dies and its strength and hardness is achieved by the actual squeezing together of the steel molecules, (FIGURE 1). The breaking strength of the wire may range from 220,000 to 250,000 pounds per square inch.

The cables of suspension bridges are made up of either wire rope, wire strands, or parallel wire. The rope or the strand is made up of a number of twisted wires, which can be stored on a reel (FIGURE 2). The strand can be cut to length, strung up as a unit and complete cable made of 19 or 37 strands can be easily erected. A few fillers and wrapping complete the cable. Strands were used for spans under 1000 feet. Today cable-stayed girder bridges that use strands are more economic for these spans.

The advantage in using strands for the cable is the ease in erection. The dubious disadvantages are the lower permitted working stresses and the lower modulus of elasticity.

Parallel wire is placed just as it says in parallel and is used on bridges of more monumental proportions and until a few years ago strung wire by wire.

In spinning the cables i.e., carrying the wires across the void, they are carried by what looks like a bicycle wheel (FIGURE 3), and ever since the days of Ellet and Roebling, it has been customary for convenience and quality to break the large parallel wire cables down into relatively small bundles of wire also called strands. When each strand is finished it is adjusted into the main body of the cable, and after placing all the strands they are squeezed by compacting into the circular form of the permanent cable, which often contains many thousand of individual parallel wires.

In several recent bridges they have erected parallel wire in bundles or strands rather than individual wires. A technique was developed to reel parallel wire strands. More will be provided on this subject by the author of the next article, Professor Konishi.

Up through the construction of the Golden Gate Bridge the practice of strand making consisted of spinning the wires in separate small saddles apart from the main cable saddle. On completion of each strand it would be checked, banded, and lowered into place in the main cable saddles, after which it would be length-adjusted to position in the main cable.

Starting with the Triborough Bridge in New York and the San Francisco-Oakland Bay Bridge, the practice of spinning the strands in place in the main cable saddle was used from time to time. Now this method is more or less standard practice.

A. I. Zuckerman is a partner in the consulting engineering firm Steinman, Boynton, Gronquist & Birdsall, 50 Broad St., New York, New York 10004.

0077–8923/80/0352–0041 $01.75/1 © 1980, NYAS

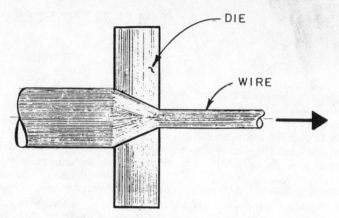

FIGURE 1. Drawing wire through die.

The tower saddles provide the support for the cables on the towers and also provide a smooth change in direction of the cables from the upward thrust of the side span to the downward plunge of the main span. The saddles are usually cast steel, but several recent saddles have been machined from weldments or a combination of casting and welding.

Depending on the erector, the cables are placed in the saddles with the strands forming a hexagon in either of two configurations, with the flat top and bottom or the flat side to side, (FIGURE 4).

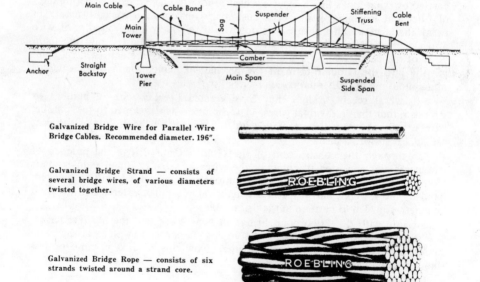

Galvanized Bridge Wire for Parallel Wire Bridge Cables. Recommended diameter. 196".

Galvanized Bridge Strand — consists of several bridge wires, of various diameters twisted together.

Galvanized Bridge Rope — consists of six strands twisted around a strand core.

FIGURE 2. Suspension bridge data.

FIGURE 3. Spinning cable.

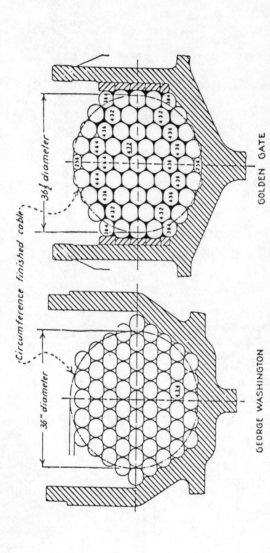

FIGURE 4. New cable make-up on the Golden Gate Bridge compared with George Washington Bridge cable. For the first time the number of wires per strand is varied to secure a more perfect circle through the saddle. The strands are also in vertical tiers instead of horizontal rows as formerly.

Using the flat side to side, called the "vertical hex," makes it possible to separate the partly completed cable into vertical tiers of strands and fix their position in the saddle both vertically and laterally during and after spinning. The base of the saddle is designed with a series of grooved rectangular steps. Permanent fillers are provided, firmly anchored against deformation, to act as retaining walls for the strands, which have to stand at least partially unsupported by adjacent strands at various stages of construction (Figure 5).

Using the flat top and bottom, called the "horizontal hex," we may use a smooth bore saddle with provisions for temporary pins to keep the strands separated. These pins are set directly into the saddle and as the spinning is advanced the pins are moved up in position, (Figure 6). The major consideration here is to keep the parallel strands isolated, in their proper place, and adjustable.

This same principle is employed in spinning the strands through the splay area.

I have used the term adjustment a number of times. Adjustment means the positioning of the small bundle of wires to its exact place in the cable. The adjustment is made at the anchorages usually by means of jacks pulling on the strands. The anchorage resists the enormous pull of the cable, and eyebars or anchor bolts to support the parallel wire strand are used to anchor the cable to the anchorage. For a strand cable the ends of the strands may be socketed and the sockets held in place with anchor bolts or through pipe sleeves and adjustment made by means of shims under the sockets, (Figures 7 and 8). In order to spread the load out over a wide area the cable is splayed or fanned out. The splay is managed by either a splay saddle or special cable band.

After spinning, the cables are compressed or compacted to a round shape, then the cable bands which support the suspenders are located and tightened on the cable with bolts. These bolts are tightened to a prescribed tension by measuring the length of the bolts before and after tightening or by measuring the tightening force. The cables tend to become more compact as loads are applied, and the cable band bolts must be checked from time to time and retightened before the structure is completed (Figure 9).

After at least 70% of the dead load is in place, the cable may be wrapped. On most of the existing bridges, soft galvanized wire was used to wrap the cable and then the wire painted (Figure 10).

To insure water tightness all the exposed portions of the saddles are covered, the cable bands caulked, and hoods used to cover the transitional shape of the cable as it emerges from the saddles. But in spite of our efforts the water may get in, and it has been our practice to omit a section of caulking from the bottom of all cable bands to allow any trapped water to drain.

Some engineers have used more modern material for wrapping the cable: fiberglass covered with an epoxy syrup or neoprene. Time will show what is the most suitable material. We have had the opportunity of exposing a number of cables by unwrapping the wrapping wire, and we have found, in some cases, that cables 60 years old seemed as though they were erected yesterday whereas others not so old were in rather poor condition.

Cable bents are usually used on suspension bridges—when the designers prefer to locate the anchorages further shoreward without increasing the length of the spans. The slopes of the cable on the anchorage side of the bent are usually steeper, which results in a greater tension on this side tending to push the cable bent riverward. It has been our practice to use an additional strand

FIGURE 5. "Vertical hex."

FIGURE 6. "Horizontal hex."

FIGURE 7.

FIGURE 8.

FIGURE 9. Checking cable band bolts.

FIGURE 10. Wrapping cable.

on the anchorage side of the cable by burying this strand in the cable bent
saddle to prevent the cable bent slipping. On the Mackinac Bridge a parallel
wire strand was wrapped around a special shoe on the cable bent saddle
(FIGURE 11).

A similar method was used on the Bosporous Bridge where the slope of the
side span cable was much steeper than the main span. The tension in the side

FIGURE 11.

span is proportionately greater requiring a greater cable area. Here four strands of wire were used around two special shoes on the tower saddle to provide the additional area (FIGURE 12).

Generally the method for spinning the cable and the details of the appurtenances are little different from those used on the Brooklyn Bridge constructed almost 100 years ago. I suppose because the suspension bridge is such a basic idea, there has been little room for improvement, except perhaps in special details. We should, however, never underestimate the ingenuity of engineers.

FIGURE 12.

LATEST DEVELOPMENTS ON PREFABRICATED PARALLEL WIRE STRAND IN JAPAN

ICHIRO KONISHI

INTRODUCTION

A number of long-span suspension bridges are being constructed under the Honshu-Shikoku (Honshi) Bridge Project in Japan. Since the completion of the Kanmon Bridge in 1973 with a main span length of 712 m, prefabricated parallel wire strands (PWS) have come to be used for bridge cable construction. In building the Kanmon Bridge much time was saved by using PWS, and an excellent cable-void ratio of 16–17% was achieved. This paper presents the latest developments and merits of PWS and includes guidelines for suspension bridge cable.

HISTORY OF SUSPENSION BRIDGE CABLE IN JAPAN

The construction of long-span suspension bridges in Japan began with the Wakato Bridge completed in 1962, the biggest of its kind in the Orient, with a main span of 367 m and cable diameter of 508 mm. The construction of this bridge used all the technology available in Japan for bridge construction at that time; however, it was not made of parallel wire cable but of spiral strand cable. A construction program was then planned for long-span suspension bridges to cross from the Japanese mainland (Honshu) to Kyushu and Shikoku Islands, which demanded much longer spans than that of the Wakato Bridge, thereby producing an incentive for the study and development of parallel wire cable.[1]

Japan's history of parallel wire cable use is not long, but together with a study of air spinning (AS) construction, engineering for the manufacture of parallel wire cable was begun to study prefabrication at shop and erection at site. Suspension bridges of small and medium size were in due course constructed, as shown in TABLE 1, by both the AS and PWS methods, which had attained respectability as field technologies by the time construction of the Kanmon Bridge was begun.

In 1968, the Kanmon Bridge (for the fixed crossing between the Japan mainland (Honshu) and Kyushu Island) was started, and parallel wire cable was chosen as the bridge cable. After a comparison study, the PWS construction method was chosen for its available merits of economy and shorter erection time. The actual construction of the bridge cable began in 1971, went on as scheduled, and was completed successfully.

For the construction of bridges to connect Honshu and Shikoku Island, the Honshu-Shikoku Bridge Authority was established in 1970 and was getting briskly under way when the so-called oil shock struck in 1973 and the construction program was suspended for a while. In 1975, construction of the Honshi Bridge Project began.

Ichiro Konishi is a Professor Emeritus of Kyoto University, Kyoto, Japan.

0077–8923/80/0352–0055 $01.75/1 © 1980, NYAS

TABLE 1

SUSPENSION BRIDGES IN JAPAN

Bridge Name	Completed in	Span Length (m)	Cables				
			Cable Construction Methods	Cable Diameter (mm)	Number of Strands	Number of Wires in Strand	Wire Diameter (mm)
Wakato	1962	89+367+89	Spiral Strand	508	61	—	—
Kanatani	1967	40+150+40	AS	114	4	108	5.00
Hakogase	1967	60+206+60	AS	143	4	168	5.00
Hachiman	1968	35+160+37	PWS	131	7	80	5.00
Kamiyoshino-gawa	1971	63.4+253.5+63.4	AS	209	7	204	5.00
			PWS		14	102	5.00
Hinoshima	1972	48.4+174+65.9	PWS	141	7	91	5.00
Kanmon	1973	178+712+178	PWS	664	154	91	5.04
Hirado	1976	105+465.4+105	AS	365	19	228	5.00
Innoshima	1982 (Schedule)	250+770+250	PWS	610	91	127	5.17
Ohnaruto	1983 (Schedule)	93+330+876+330	PWS	830	154	127	5.37
North Bisan-seto	1987 (Schedule)	270+990+270	PWS	—	—	127	—
South Bisan-seto	1987 (Schedule)	270+1100+270	PWS	—	—	127	—
Shimotsui-seto	1987 (Program)	230+920+230	AS	—	—	—	—

The Honshi Bridge Project [2] is to construct many bridges over three routes, as shown in FIGURE 1. The length of each route, structural standard and construction costs are shown briefly in TABLE 2. Besides the arch-type Ohmishima Bridge already completed, with a 297 m span, five suspension bridges are planned for the first phase of construction under the Honshi Bridge Project. Those under construction are the Innoshima Bridge, set for completion in 1982, the Ohnaruto Bridge to be completed in 1983, the North Bisan-Seto Bridge set for completion in 1987, and the South Bisan-Seto Bridge also set for completion in 1987. In addition, the Shimotsui-Seto Bridge is now intended for completion in 1987. (See FIGURE 2)

These bridges, with the exception of the Shimotsui-Seto Bridge, which is programmed for use of the AS method to keep up development of the AS technique of making suspension bridge cable for possible future needs, are scheduled for the PWS bridge cable construction method. Among these bridges, the South Bisan-Seto Bridge will have a maximum span of 1100 m length with the cable diameter of 1 m or so.

The Innoshima Bridge, the first suspension bridge to be built under the Honshi Bridge Project, has PWS of 1360 m length with 127 wires, larger than that of the Kanmon Bridge. At present, 182 strands of 127-wire PWS are being manufactured for the Innoshima Bridge, and 308 strands of 127-wire PWS will follow for the Ohnaruto Bridge.

As the result of a series of actual constructions under big projects such as the Kanmon Bridge and the Honshi bridges, bridge parallel wire cable construction, especially by the PWS method, is developing remarkably.

BRIEF EXPLANATION OF THE MANUFACTURE OF PWS

A brief explanation is given here of the important points of PWS manufacture in Japan.

The manufacturing line consists of, as shown from the left in FIGURE 3, a series of bobbins to supply wire, screens to arrange and bundle wires, forming rollers to make wire into a strand with a hexagonal cross section, a seizing machine to automatically seize by plastic tape the hexagonal bundle of wire at regular intervals, and a reeling machine. FIGURE 4 shows a strand being wound by a reeling machine.

Wire length is strictly controlled in the manufacture both of wire rod and galvanized wire to avoid splices and improve productivity. Strand is reeled up on a steel reel 1.8 m in diameter. One of the points of reeling is to allow a strand to reel freely so that the strand may twist, without being forced to twist.

Each strand is cut to a given length and socketed using an alloy of Zn-Cu (Cu 2%), which is melted and poured. This alloy is adopted because its resistance against creep is higher than that of any other pouring metal. The socket is so designed that stress worked on the surface contacting the poured metal is below 4 kg/mm^2.

Determination of the strand length is obtained by setting a wire called a "gauge wire," which has been exactly measured for a given length in advance, at the apex position of a hexagon. Marks on the gauge wire are transferred onto a strand for the determination of the strand length. The length of the gauge wire itself for PWS in the case of the Kanmon Bridge was determined by extending a wire of exactly 1160 m in total length with a given tension on

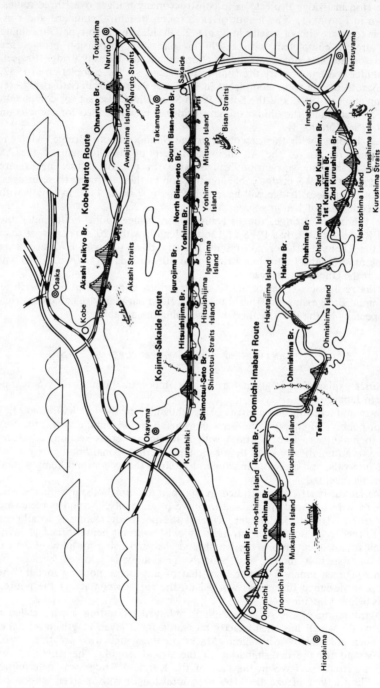

FIGURE 1. Bird's-eye drawing of the Honshu-Shikoku bridges.

TABLE 2

LENGTH, STRUCTURAL STANDARDS, AND CONSTRUCTION COSTS

Item	Category	Particulars	Kobe-Naruto Route	Kojima-Sakaide Route	Onomichi-Imabari Route
Length (km)	Highway	—	81.1	37.8	60.1
	Railway	—	89.8	49.2	—
Structural Standards	Highway	Classification	Expressway	Expressway	Expressway
		Design Speed (km/h)	100	100	80
		Number of Lanes	6	4	4
	Railway	Classification	Shinkansen	Ordinary Line and Shinkansen	—
		Number of Tracks	2	2+2	—
Construction Cost (billion yen) (1977)			1,150	840	410

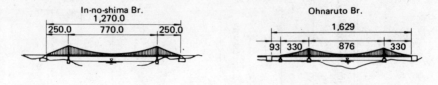

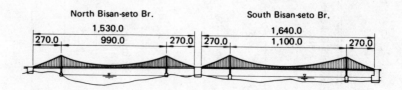

FIGURE 2. General view of the five suspension bridges scheduled for completion as the early stage of the Honshi Bridge Project.

flat straight ground and putting a mark at each point on that wire to obtain, once and for all, a so-called standard (mother) wire. The marks on the mother are in turn transferred on to another wire to make it then into the gauge wire. The gauge wire of PWS for the Innoshima Bridge was produced likewise. For the Ohnaruto Bridge, on the other hand, it was decided to obtain the gauge wire by means of a piece of magnetic measuring equipment newly developed so that effective measurements could be made conveniently in the shop.

In the manufacturing stage it is important to ensure that the lengths of the individual wires of a strand are accurate. In the case of AS, as the sag of wire is individually adjusted at the site, the difference in stress due to the difference in length of individual wires falls within the allowable range. In the case of

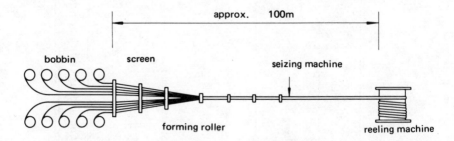

FIGURE 3. An outline of the manufacturing line for parallel wire strand.

FIGURE 4. Strand reeled on steel reel.

PWS, however, as wires are prefabricated by being cut to a given length and socketed into a strand at the shop, any difference in length between individual wires causes a difference in the tensile stress of the individual wires of a strand after erection. Experiments so far performed ascertained that the length dispersion in the manufacture of individual wires of PWS was below 0.01% of the length, which presents an absolute value of 2 kg/mm² in terms of the difference in stress among wires. Difference in stress among wires of a strand produced by saddle curvature, determined by calculation, was found to be within the allowable range.

CONSTRUCTION OF PWS CABLE

FIGURES 5 and 6 show a diagram and picture of the Kanmon Bridge. Construction of the bridge cable was accomplished within the scheduled period of two months from September through November, 1971, for the total 154 strands of 91-wire PWS, 1160 m long for each cable.

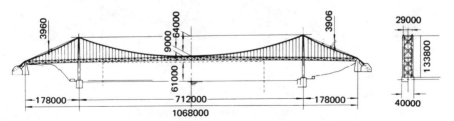

FIGURE 5. General view of the Kanmon Bridge.

FIGURE 6. The Kanmon Bridge.

A reel set up on an unreeling machine at an anchorage paid out the strands along rollers arranged on the catwalk by means of a hauling rope 32 mm in diameter that was driven by a 160 HP remote control hauling engine with the tension being controlled, as shown in FIGURES 7 and 8. The hauling velocity was 60 m/min at the maximum at the center span and 40 m/min on average. The average construction per day was 4 strands of each cable, making 8 strands altogether. After being hauled, the strands were transferred from the rollers to a given place on the saddle.

Strands are generally arranged into a hexagonally shaped cable section because a hexagon gives a smaller section for the cable. There are two ways of making a hexagonal arrangement [3]: either the apex is at the top and bottom

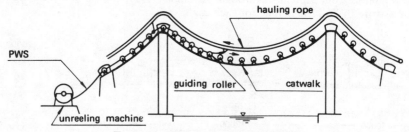

FIGURE 7. PWS construction method.

or the side lies at the top and bottom. The former arrangement is represented, for instance, by the Golden Gate Bridge and the Forth Road Bridge, while the latter was used in the George Washington Bridge and the Verrazano-Narrows Bridge, though those were constructed by the AS method. The Kanmon Bridge, where the strands are laid in a rectangular shape in a groove on the saddle to obtain a lower cable-void ratio, adopted the first arrangement (see FIGURE 9) for these reasons:

1. By using this method a spacer can easily be installed vertically to prevent adjoining strands in a saddle from interfering with each other.

FIGURE 8. Strand hauled on rollers on the catwalk.

2. The cable-former fitted at several points on the catwalk to arrange the strands in a consistent position can be designed in the form of a comb, which also enhances the thermal adaptability of the cable to the ambient temperature.

Because of the good experience with this strand arrangement in the Kanmon Bridge, all the cable arrangements of bridges under the Honshi Bridge Project were designed with a vertical hexagon apex.

The first strand to be erected becomes the standard strand for sag adjustment of all later strands. The standard strand was adjusted to lie exactly as measured in the designed sag position. This operation could be carried out only on a night when the temperature remained constant across the entire span and when the wind velocity was under 14–16 m/sec. Strands subject to sag adjustment were set preliminarily a little above other strands already adjusted. The strands were firmly fastened on one tower and sag for the center span was adjusted first. Then an adjustment of the sag was done in turn for the side span, for the back stay and at the front of the anchorage by inserting shim plates. The total shim plate volume inserted for the Kanmon Bridge for both socket ends, as shown in FIGURE 10, was 89.5 mm for all strands on an average and 170 mm at the maximum. Taking this value for reference in calculating the manufacturing accuracy of each strand, the length dispersion remained well within ±80 mm and so therefore within a range of ±0.007% in ratio for entire length of 1160 m.

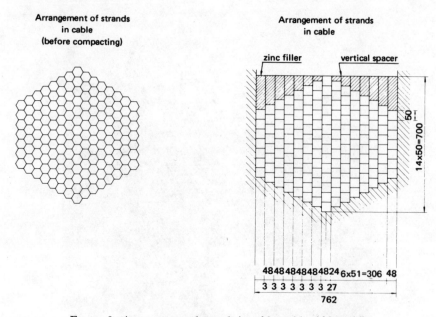

FIGURE 9. Arrangement of strands in cable and in cable saddle.

The value ±0.007% in dispersion includes any errors in erecting the cable, so the strand accuracy by itself in the manufacture may be considered even better.

In order to decide cable diameter and to design cable bands for the Kanmon Bridge, a cable model was made up in advance with 154 strands, each with 91 wires, 10 m long with the same composition as the real wires. By means of a 600-ton hydraulic type compactor a study was carried out concerning the relation between oil pressure and cable diameter, the compacting interval in the longitudinal direction, the finally obtainable cable diameter and the cable-void ratio. Based on these experiments, the compacting was carried out at intervals of 1 m using a hydraulic compactor with a 600-ton capacity (100 ton × 6 cylinders).

The resulting void ratio was 16–17% as shown in TABLE 3, a lower value

TABLE 3

ACTUAL CABLE DIAMETER AND VOID RATIO ON THE KANMON BRIDGE

Cables	Measured Points	At Cable Bands						Between Cable Bands	
		Cable Diameter (mm)			Average Area of Filler (cm²)	Approx. Area of Cable (cm²)	Void Ratio (%)	Cable Diameter Average (mm)	Void Ratio (%)
		Vertical	Horizontal	Average					
East Cable	1 *	656	664	660	1.0	3420	18.0	669.5	20.5
	2	654	661	657.5	0.8	3392	17.4	669.5	20.5
	3	654	655	654.5	6.0	3363	16.6	667	19.8
	4	653	650	651.5	34.9	3303	15.1	658	17.8
West Cable	1	655	663	659	1.2	3409	17.8	667	19.8
	2	655	660	657.5	1.2	3399	17.6	666.5	19.8
	3	654	656	655	6.7	3363	16.6	664.5	19.3
	4	652	651	651.5	34.9	3303	15.1	657.5	17.6
Average	East Cable	654	658	656	7.9	3372	16.8	667	19.8
	West Cable	654	658	656.5	8.4	3371	16.8	665	19.3
	Overall	654	658	656	8.2	3372	16.8	666	19.5

* 1: Shimonoseki side span; 2: Center span (Shimonoseki side); 3: Center span (Moji side); 4: Moji side span.

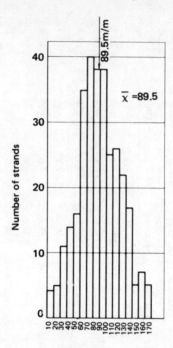

Variation of shim volume (mm)

FIGURE 10. Histogram of shim volume.

than ever previously observed in parallel wire for other bridges. The lower void ratio means a superior parallel wire cable with fewer intersections.

Traditional wire wrapping was adopted in the Kanmon Bridge. The wrapping wire was made of galvanized steel 4 mm in diameter. The wire tension when winding was 150 kg. A paste of highly polymerized lead was applied on the cable surface before wrapping to obtain greater corrosion resistance. The wire wrapping method is planned for the bridges of the Honshi Bridge Project; however, for cable weather protection another method such as plastic covering or something similar is being considered.

PRACTICAL WORKABILITY TEST

There had been some small and medium-length suspension bridges built with PWS, but the Kanmon Bridge required bridge cable strands such as had never been made before—91-wire PWS and 1160 m long. Therefore, unreeling and hauling tests of PWS were conducted in 1970 to clarify issues on strand manufacture and erection, using 91-wire PWS, 1160 m long, unreeled for the entire length on rollers set at regular intervals on flat straight ground over a

1.2-km length. Wires thus laid out in the experiment were used afterwards as the gauge wire for the actual cable strand.

Strand larger than that of the Kanmon Bridge was studied for the Innoshima Bridge, the first of the Honshi Bridge Project suspension bridges, with the intention of increasing effectiveness both in manufacture and in construction. It was then decided that the cable composition would be 91 strands, each with 127-wire PWS 1360 m long. Taking into consideration the decision thus made and the future possible requirements for longer strands, more unreeling and hauling tests were conducted in 1977 using PWS of 127-wire of 5.17 mm diameter, 1360 m long and 30.7 tons in weight extended likewise for the entire length on rollers set at regular intervals on flat straight ground over a distance of 1.4 km to ascertain the adequacy for the field construction on items such as:

1. Strand loose condition in reel;
2. Longitudinal slippage of wire in the strand after hauling;
3. Twisting and rotation of the strand;
4. Occurrence of the "bird-cage" phenomenon;
5. Damage of seizing tapes;
6. Workability of unreeling machine and other equipment;
7. Hauling velocity of strand and measurement of hauling tension.

No phenomenon detrimental to the actual construction was observed, and 127-wire PWS of the 1400 m long class was confirmed as being the most satisfactory for adoption.

FIGURES 11 and 12 show the unreeling and hauling tests. They will occur in the summer of 1980 when the cable erection for the Innoshima Bridge will

FIGURE 11. Unreeling of strand.

begin. The experiment described above indicates good prospects for the field erection.

The comparative bridge cable specifications are given in TABLE 4 to show a transit of development from the Kanmon Bridge, to the Innoshima, and Ohnaruto Bridges. Weight per reel is markedly increased for the increased number of wires and increased length of strands.

FIGURE 12. Strand hauled on rollers.

CONCLUSION

Japan, despite its short history of parallel wire cable, is making great progress in the PWS construction method as big projects such as the Kanmon Bridge and the Honshi bridges are developed. The main features of the PWS construction method are its unique merits of saving construction time, reducing the man days of the construction period, and effective improvement of working efficiency against wind during construction. The PWS method of bridge cable construction is on the way to real development in Japan in line with the AS construction method.

The hope for the future is that PWS will be used for longer and larger bridge construction.

TABLE 4

COMPARISON OF THE BRIDGE CABLE DIMENSIONS

Item	Kanmon	Innoshima	Ohonaruto
Cable Diameter (mm)	664	610	830
Number of Strands	154	91	154
Number of Wires in Strand	91	127	127
Wire Diameter (mm)	5.04	5.17	5.37
Unit Weight of Strand (kg/m)	14.2	20.9	22.5
Strand Length (m)	1162	1360	1722
Gross Weight (ton) (including reel)	17.5	30.7	41.0
Section of PWS (mm)	48.7 / 55.4	58.9 / 67.2	61.2 / 69.8

REFERENCES

1. HIGO, H. & Y. NAKAJIMA. 1966. Study on the parallel wire cable of long span suspension bridges. International Symposium on Suspension Bridges :301–305 Laboratorio Nacional de Engenharia Civil. Lisbon.
2. KAWASAKI, I. 1979. Honshu-Shikoku Bridge planning in Japan. IABSE Symposium, Zurich. Symposium report, Vol. Band **32:**87–98.
3. BIRDSALL, B. 1971. Main cable of Newport Suspension Bridge. J. Struct. Division. ASCE. Vol. **97.** No. ST. Proc. Paper 8606: 2825–2835.

CABLE-STAYED BRIDGES: A CURRENT REVIEW

WALTER PODOLNY, JR.

INTRODUCTION

With increasing economic, aesthetic, environmental, ecological, energy concerns and the replacement of deficient or obsolete bridges, the next decade will certainly be one requiring innovative designs and concepts. Present and future generations of the cable-stayed bridge are certainly destined to play a major role in this future.

The cable-stayed bridge is an innovative bridge structure that is both old and yet new in concept. It is old in the sense that it has been evolving over a period of 400 years and new in that its modern day implementation began in the 1950s and within the last decade has attracted the attention of bridge engineers in the United States.

As engineers, we are aware that no particular concept or bridge type is suitable for all environments or is in itself a panacea to all considerations or problems. The selection of the proper type of bridge for a particular site with a given set of circumstances must take into account many parameters. The process of weighting and evaluating these parameters for various types of bridges under consideration is certainly more an art than a science. This paper will attempt to provide data and information that will allow an intelligent assessment and evaluation of cable-stayed bridges with other types. The obvious first question to be asked is what is a cable-stayed bridge? Therefore, the cable-stayed bridge must be defined or classified.

CLASSIFICATION

Bridges that depend upon high-strength steel cables as major structural elements may be classified as suspension bridges—specifically as cable-suspended or cable-stayed bridges. The fundamental difference is in the manner in which the bridge deck is supported by the cables.

In the cable-suspended bridge the deck is supported at relatively short intervals by vertical hangers which are in turn suspended from a main cable. The main cables are relatively flexible and thus take a shape that is a function of the magnitude and position of loading. A typical example is the classical catenary suspension bridge. No one will deny the graceful beauty of the silhouette of cables and deck against the sky. In this type of structure even the uninitiated layman can appreciate that form follows function.

The inclined cables of the cable-stayed bridge support the bridge deck directly with relatively taut cables which, compared to the classical suspension bridge, provide relatively inflexible supports at several points in the span. This type of structure also makes it easy to see form following function.

Walter Podolny, Jr. is with the Bridge Division, Office of Engineering of the Federal Highway Administration, United States Department of Transportation, Washington, D.C. 20590.

0077–8923/80/0352–0071 $01.75/1 © 1980, NYAS

At this point the cable-stayed bridge has been defined or classified. The obvious next question is why should a cable-stayed bridge be considered?

WHY CABLE-STAY?

To answer this question, the Luling Bridge in Louisiana crossing the Mississippi River will be used as an example. Why was a cable-stayed system selected for this crossing? Certainly there are other structural systems suitable for the given span range that are more familiar to builders of bridges across the Mississippi River. Foremost in this respect is the venerable cantilever truss, so very common on the Mississippi below St. Louis. However, the cantilever truss and its variations are heavy on steel usage for spans over 1,000 ft (300 m) in length. Moreover, maintenance and inspection costs of long-span truss bridges are high because of their numerous structural components, many of which are not easily accessible.

In contrast, a cable-stayed system employing deck girders for stiffening of the bridge floor offers immediate advantages in terms of weight alone. The reason for this being so is clearly evident in FIGURE 1.[2] It shows a comparison of bending moments in a five-span unstayed design and a cable-stayed structure selected for the Luling Bridge. The ratio of these moments is close to 1:10 in favor of the stayed system. Moreover, these moments can be controlled to make them more evenly distributed along the structure. In the process, material utilization is more efficient, even with a very low depth-to-span ratio of 1:90

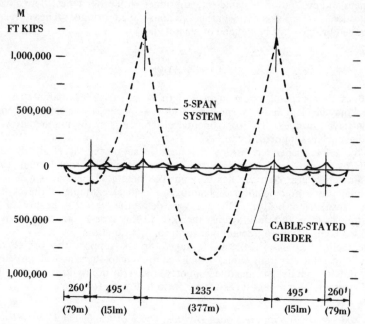

FIGURE 1. Moment diagram comparison, cable-stayed girder vs. five-span system. (Courtesy of Mr. Stanley E. Jarosz)

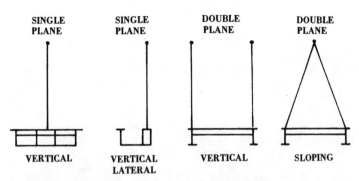

FIGURE 2. Transverse cable arrangement.

as is the case of the Luling Bridge. This factor and the anticipated low cost of maintenance, as compared with trusses, had a significant effect on the selection of a cable-stayed system.

SPAN PROPORTIONS

In any type of bridge structure one of the first design considerations to be evaluated is the span proportions. In cable-stayed bridges the span arrangements are of three basic types: two spans, symmetric or asymmetric; three spans; or multiple spans. A partial survey of two-span asymmetric structures indicates that the longer span is in the range of 60 to 70 percent of the total length. Exceptions are the Batman Bridge in Australia and Bratislava Bridge in Czechoslovakia, which have a ratio of 80 percent. However, these two structures do not have the back-stays distributed along the short side span; they are concentrated into a single back-stay anchored to the abutment. A similar survey of three-span symmetric structures indicates that the ratio of center span to total length is approximately 55 percent. In multiple-span structures the spans are normally of equal length, with the exception of flanking spans that connect to approach spans or abutments.[3]

GEOMETRIC CONFIGURATION OF STAYS

A wide variety of stay geometry has been employed in the construction of cable-stayed bridge systems. The arrangement of stays, girders, and pylons is subject to highway requirements, site conditions, and aesthetic preferances.

In a direction transverse to the longitudinal axis of the bridge, two basic arrangements have been used for the stay arrangement: the stays are arranged in either two planes or in a single plane (FIGURE 2). Each of these two basic arrangements have variations as indicated.[3] A three-plane system is illustrated in FIGURE 3 by a third-prize winner in the Danish Great Belt Bridge Competition. Design requirements were for three lanes of vehicular traffic and single rail traffic in each direction. The solution proposed by the English consulting firm of White, Young and Partners contemplated three vertical stay planes, one in the median and one along each exterior edge of the roadway deck. This may

FIGURE 3. Danish Great Belt Competition. (Courtesy of White Young & Partners).

become a more important concept in the future, especially in congested urban areas where requirements, in some instances, are for three or more lanes in each direction and may also incorporate preferential vehicle lanes, car pool lanes, or mass transit requirements.

The configuration of stay geometry, in elevation, is only relevant where there is more than one forestay or backstay. In elevation, there are four basic stay configurations that are amenable to the single and double plane systems previously described. First, there is the radiating or converging type, where all the stays are taken to the top of the pylon. Second is the harp system, where the stays are parallel to each other, being equally spaced along the girder and the pylon. Third, there is the fan type, where the cables are equally spaced along the girder and pylon but are not parallel to each other, a hybrid of the

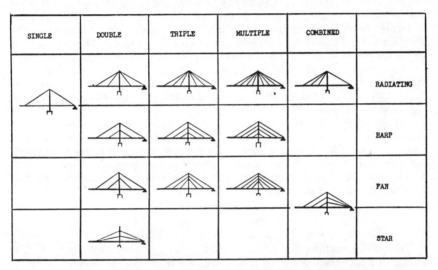

FIGURE 4. Geometry matrix—stay configuration in elevation.

first two types. The fourth and last basic type is the star configuration, where the stays are spaced along the pylon and converge to a common point on the girder.

The fan configuration represents a compromise between the extremes of the radiating and harp systems. It generally evolves from a radiating system when the accommodation of the forces at the top of the pylon becomes too complex. The star system is perhaps defensible from an aesthetic point of view; however, the convergence of the stays at the deck level is in direct opposition to the view that the stays should be distributed as much as possible at the girder.

A tabular summary of the various stay arrangements is presented in FIGURE 4.[3] The tabulation is a matrix of the four basic stay types and the number of stays emanating from one side of the pylon. It can be seen that many other variations and combinations are possible. Therefore, the cable-stayed bridge is not simply one bridge type, but many different individual types evolving from an extremely versatile concept of bridge design. It is important not to restrict a type selection study to only one cable-stayed type.

Pylons

There are numerous possible pylon arrangements. They may be single, twin, portal frame or A-frame.[4] They may also be vertical or inclined. Pylons may be hinged or fixed at the base. Although larger bending moments are produced in the pylon when fixity is provided, most of the cable-stayed bridges have been built with fixed-base pylons. The advantage of increased rigidity of the structure obtained with fixed bases, especially during erection, offsets the disadvantage of higher bending moments.

The behavior of the pylon will depend upon the details of its connection to the stays, deck and substructure. In addition to its own weight, the pylon carries a substantial portion of the total weight of the structure transmitted by the stays. Pylons should be designed as members subjected to axial compression and combined bending moment from both directions. The stability of plate elements should be investigated. In addition, overall pylon buckling in the longitudinal and transverse direction should also be evaluated. Overall buckling in the longitudinal direction is a function of the stiffness of the stays, primarily the back stay. The effective overall buckling length may be determined by a stability analysis of the entire system. The stiffness of any bracing that may be present should be considered in determining the effective buckling length in the transverse direction.

The height of the pylon in relation to the span will obviously have an influence on the cable forces and thus the amount of cable steel required (Figure 5).[5] The ratio of pylon height above the bridge deck to center span length for a three-span structure should preferably be in the range of 0.16 to 0.2. For a two-span asymmetric structure, the longer span may be considered as one-half the span of a three-span symmetric structure.

Deck Structure

The deck structure may be constructed of steel or reinforced concrete. The choice of materials is primarily a function of availability and economics prevalent at a particular time and specific location of the bridge site. Basically there are two types of girders, the stiffening truss type for steel construction and the solid web type used for both steel and concrete. In both steel and concrete construction, girders or box girders or a combination of the two may be used. In addition, concrete can also be cast-in-place or precast, and segmental construction may also be considered. With the exception of conventional catenary-type suspension bridges, the stiffening girder is seldom used today. Trusses, as compared to solid web girders, require more fabrication and are relatively harder to maintain, corrosion protection is difficult, and trusses are aesthetically unfavorable.

A survey of 20 cable-stayed bridges indicates a depth-to-span ratio of the girder varying from 1/40 to 1/80. In a multistay system, such as the Pasco-Kennewick Bridge, the depth to span ratio is 1/140.

For the bridge deck of the superstructure it is important to design the floor system in such a manner that the whole deck plate can participate in carrying the large longitudinal normal force in composite action with the main girders. Illustrated at the top of Figure 6 is the old type of superstructure where the deck, stringers, cross beams, lateral bracing, and main girders act independently. At the bottom of Figure 6 is the new type of deck in which the orthotropic

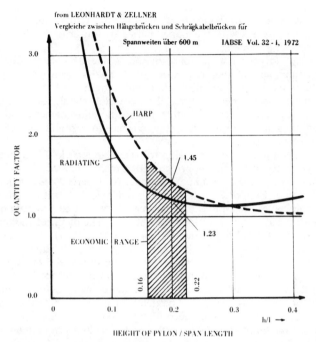

FIGURE 5. Stay cable steel quantity as a ratio of pylon height above deck to span length. (Courtesy of F. Leonhardt)

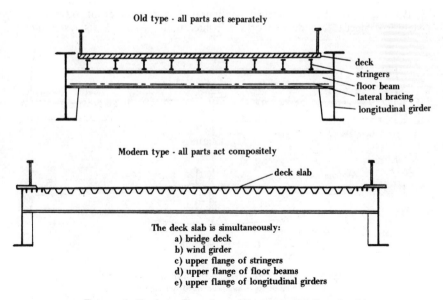

FIGURE 6. Deck configuration. (Courtesy of F. Leonhardt)

plate is simultaneously the bridge deck, lateral wind girder, upper flange of the stringers if they are present, the upper flange of the cross beams, and upper flange of the main girders.[4]

The normal forces or axial force in the bridge deck is produced by the horizontal component of the cable force at its anchorage with the girder. In steel structures the orthotropic deck plate with longitudinal stiffeners is a good solution. In concrete superstructures the floor slab should be spanned longitudinally between cross girders so that the main reinforcement will be longitudinal and thus assist in resisting the normal forces and to reduce creep deformations.[4]

Leonhardt,[4] in a competitive design for the Rheinbrücke-Flehe near Düsseldorf, an asymmetric cable-stayed bridge with a major span of 1,214 ft (370 m) and multistays, designed an economical and easy to erect superstructure, indicated at the top of FIGURE 7. It is a steel orthotropic deck with single web edge beams for the cable anchorages and two centrally located longitudinal girders for the erection derricks and for load distribution. The depth of the longitudinal girders is only 5.6 ft (1.7 m) at the edge and 7.9 ft (2.4 m) at the center. It is shallow because of the stiffness provided by the many cable stays. The width of the structure is 135 ft (41 m). Almost the same cross section can be accomplished with a concrete deck, shown at the bottom of FIGURE 7. The concrete deck, between concrete edge beams, spans longitudinally between and is composite with the steel cross girders. The depth of the composite concrete section at the center is 11.5 ft (3.5 m). According to Leonhardt, this type of deck can be economical for three-span cable-stayed bridges with a center span of 2,300 ft (700 m).

STAY-CABLE SPACING AND ANCHORAGES

The choice of the geometrical configuration and number of stays in the system is subject to a wide variety of considerations. Few cable stays result in

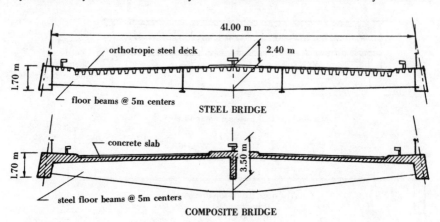

NOTE: no bottom flange required for both sections
spacing of stays independent of spacing of
floor beams

FIGURE 7. Modern orthotropic and concrete deck sections. (Courtesy of F. Leonhardt)

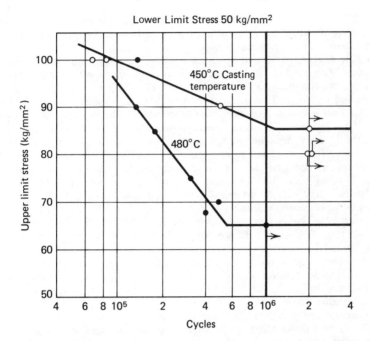

FIGURE 8. Fatigue test of wire 5 mm diam. with zinc-copper alloy-filled sockets. (Courtesy of Der Stahlbau [7])

large stay forces which, in turn, require massive and complicated anchorage systems and consequently, reinforcing of the bridge girders to transfer shear, moment and axial loads. A relatively deep girder is required to span the large distance between stays. Depending upon the location of longitudinal main girders with respect to the cable-stay planes, large transverse cross girders may be required to transfer the stay forces to the main girders. A large number of cable stays, approaching a continuous supporting elastic media, simplifies the anchorage and distribution of forces to the girder and permits the use of a shallower depth girder.[3, 6]

Cable-stayed bridges in Germany started with 3 to 5 stays at each side of the pylon, but have gradually evolved to more stays at shorter spacing to simplify the anchorage detail and erection. The construction of the deck can be simple and economical if the spacing of the stays is such that the deck can be erected road-way width by free cantilever methods from stay to stay without auxiliary methods. The depth of the roadway girder can be kept at a minimum, the deck becomes more or less the bottom chord of a large cantilevering truss and needs almost no bending stiffness because the inclined stays do not allow any large deflections under concentrated loads.[4]

With the helical strand type of cable, the wires at the end of the strand are carefully placed in the basket of the socket, which is then filled with molten zinc. The Japanese have reported [7] that the pouring temperature of the zinc alloy when filling the socket affects the fatigue strength of the wires at the socket (FIGURE 8). A casting temperature of 450° C results in decreased fatigue strength and fractures in, or in the vicinity of, the cable anchorage. There have been some advancements in the technology of cable fittings.

The BBR type of terminal hardware as used in prestressed, post-tensioned concrete construction has been utilized with button-headed parallel wire cables in some cable-stayed bridge applications.

A new anchorage system (FIGURE 9) developed by Dyckerhoff and Widmann has been utilized for the first time, in conjunction with parallel Dywidag bar stays, on the Main River Bridge near Hoechst, a suburb of Frankfurt, constructed in 1971. This anchorage consists of 25 threaded ⅝ in. (16 mm) diameter bars encased in a steel tube or conduit. The cross sectional orientation of the individual bars are maintained by a polyethylene spacer. In the intermediate or dynamic anchorage zone, an end-piece connected to the conduit has rivets on its surface to improve bond when it is embedded in the construction concrete. This end-piece is closed with a guide cap from which the individual bars in individual conduits lead to end anchorages. The Dywidag parallel bar stay offers, among other advantages, favorable behavior with respect to wind-excited vibrations.

A new type socket has been developed by BBR Ltd of Zurich, Switzerland (FIGURE 10). It is named HiAm for *High Am*plitude socket and is marketed in the United States by INRYCO, a licensee of BBR. A similar socket is also available from the Prescon Corp. The HiAm socket has shown improved results in tensile, fatigue, and creep tests. It utilizes a casting material composed of steel balls, zinc dust, and an epoxy resin mixture. This assembly consists of parallel wires of ¼ in. (7 mm) diameter with button heads bearing on a steel anchorage plate, which closes the end of the conical cavity in the steel socket, the void being filled by the special mixture of steel balls, zinc dust, and epoxy resin, which ensures a gradual transfer of the cable force. This socket has been successfully fatigue tested for 2 million cycles with a stress range of 35 to 45 percent of ultimate strength of the wire.

ECONOMIC EVALUATIONS

Limited experience to date indicates that cable-stayed bridges with principal spans less than 500 ft (150 m) are most suitable for pedestrian bridges. However, under certain site conditions, cable-stayed bridges with spans under 500 ft (150 m) may still be a viable solution, but the deciding parameter is usually not an economic one.

In the case of the 450 ft (137 m) center span Sitka Harbor Bridge, six different types of bridges were evaluated before a final decision was made to adopt the cable-stayed system (TABLE 1). With the exception of the first two types, the most that can be said is that the cable-stayed system is competitive, but there is no clear-cut economic advantage. That is to say, economics is not a dominant parameter in the decision to select the cable-stayed type structure.

In the intermediate span range, a study by P. R. Taylor in Canada[8] (FIGURE 11), compared girder bridges, suspension bridges, and cable-stayed bridges using an orthotropic steel superstructure, and concluded that with center spans ranging from 700 to 800 ft (210 to 240 m) the cable-stayed bridge was most economical by a 5 to 10 percent margin over other types of comparable bridges.

The economic evaluation of the Pasco-Kennewick Intercity Bridge studied in detail the five alternative structural designs considered to be the most feasible. Alternative 1 was a steel plate girder consisting of eight continuous spans. The superstructure consisted of four lines of girders 20 ft (6 m) on center with a

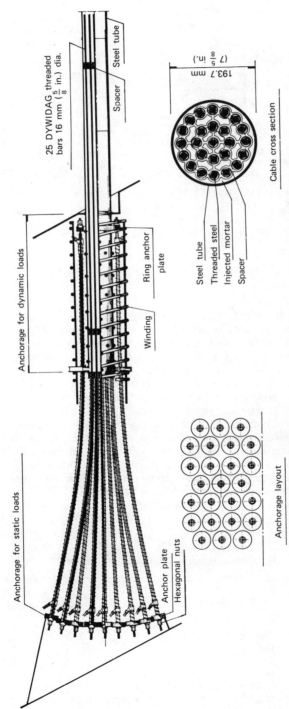

FIGURE 9. Anchorage for Main River Bridge. (Courtesy of Dyckerhoff & Widmann)

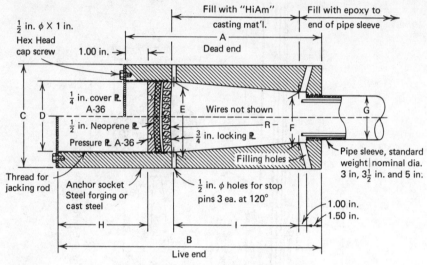

FIGURE 10. HiAm socket.

constant depth of 15 ft (4.5 m). Alternative 2 was the two-plane, three-span, cable-stayed version continuous with the approach spans. Alternative 3 proposed a prestressed concrete, single cell, incrementally launched, segmental girder of a constant 15 ft (4.5 m) depth and five interior spans of 374 ft (114 m). Alternative 4 was a post-tensioned, balanced cantilever, with segmental construction consisting of a main span of 450 ft (152 m). Alternative 5 consisted of an asymmetric cable-stayed steel box girder with a composite deck and concrete girder approach spans.

The economic comparison of these five alternatives using alternative 2, the final design choice, as a base is presented in TABLE 2. As can be seen, the estimated construction cost comparison, including substructure, shows no conclusive economic argument for the approval of any one design. Therefore, satisfactory functional requirements, anticipated long-term performance, construction and design requirements, as well as the estimated initial costs must also be evaluated in reaching a decision.

TABLE 1

SITKA HARBOR BRIDGE—COST STUDY

Type	Description	Cost Ratio (Cable-stayed Girder = 1.00)
I	Plate girder with fenders	1.15
II	Plate girder continuous	1.13
III	Orthotropic box girder	1.04
IV	Through-tied arch	1.04
V	Half through-tied arch	1.06
VI	Cable-stayed box girder	1.00

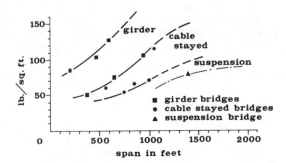

FIGURE 11. Weight of structural steel in lb/sq ft. of deck for orthotropic steel bridges (Courtesy of *Engineering Journal* (Canada)[8])

Obviously, consideration of these items must take into account the particular site conditions, local environmental conditions relative to natural hazards along with the local and national economic environment at the time the estimate is made, as well as any short- and long-term economic conditions that may affect the final cost.

Let us now look at long-span structures. The economic survey by Taylor,[8] shown in FIGURE 11, has a reference point for a cable-stayed bridge which is higher than one might expect for that magnitude of center span. It appears that this singular point is apparently based on the data taken from the Kniebrücke at Düsseldorf, which is an asymmetric bridge with one pylon. The data presented by Taylor are for a center span of 1,050 ft (320 m) with a corresponding weight of deck structural steel of 115 psf (561 kg/m²).

If these data were considered to be one-half of a symmetric two-pylon arrangement with a center span of approximately 2,000 ft (610 m) and re-plotted against the previous data, as shown in FIGURE 12, a different conclusion may be drawn. The cable-stayed bridge is then seen to compete favorably with the suspension bridge of comparable span. From this limited study it appears

TABLE 2

PASCO-KENNEWICK BRIDGE—ECONOMIC COMPARISON

Alternative	Description	Cost Ratio
1	Steel plate girder	1.005
2	Cable-stayed concrete box girder	1.000
3	Concrete box girder—push out method	0.952
4	Concrete box girder—cantilever method	0.981
5	Cable-stayed steel box girder	1.019

reasonable to assume optimistically that cable-stayed bridges may penetrate the complete range of spans now dominated by conventional suspension bridges.

Leonhardt has concluded, from studies he has conducted, that cable-stayed bridges are particularly suited for spans in excess of 2,000 ft (600 m) and may even be constructed with spans of more than 5,000 ft (1,500 m).[4]

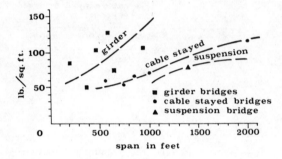

FIGURE 12. Weight of structural steel in lb/sq ft. of deck for orthotropic steel bridges.

CLOSING REMARKS

There are numerous parameters that go into the process of determining an acceptable bridge type to suit a specific site. The process of weighting and evaluating these parameters is certainly more an art than a science. The obvious question is why should a cable-stayed bridge be considered? In answer to this question it is important to recognize that the girder is supported from above by soft supports of the stay attachment rather than from below by hard supports of additonal piers. Thus, expensive piers and foundations that might have to be constructed in deep water or high expensive piers built in deep valleys can be avoided or minimized. Besides the obvious aesthetic advantages, depending on the site, there may be advantages in the elimination of concern regarding such items as horizontal navigation clearance and pier impact from run-away barges or shipping, scour problems of piers in water, ice jams, impact from flood debris or ice floes, minimization of impact on wet lands and environment, etc.

As has been indicated, the cable-stayed bridge type is one that is certainly competitive with other type structures up to 1,000 ft (300 m) spans and possibly larger spans. There is certainly the potential for this type structure to become economically favorable in the long-span range dominated by the suspension bridge.

As designers, we should be aware that the cable-stayed bridge is not simply one-bridge type, but many different individual types evolving from an extremely versatile concept of bridge design. In selecting a bridge type, the bridge engineer should not limit his thinking to just *a* cable-stayed bridge; he should consider a number of possible geometric variations of cable-stayed bridges.

REFERENCES

1. PODOLNY, W. JR. & J. F. FLEMING. 1973. Cable-Stayed Bridges—Simple Plane Static Analysis. Highway Focus. 5 (2): 46–67. Federal Highway Administration. Washington, D.C.
2. JAROSZ, S. E. 1975. Design Aspects of Luling Bridge. Paper presented at ASCE National Structural Engineering Convention, New Orleans, Louisiana.
3. PODOLNY, W., JR. & J. B. SCALZI. 1976. Construction and Design of Cable-Stayed Bridges. John Wiley & Sons, Inc. New York.
4. LEONHARDT, F. 1974. Latest Developments of Cable-Stayed Bridges for Long Spans. Saertryk af Bygningsstatiske Meddelelser. 45 (4). Denmark.
5. LEONHARDT, F. & W. ZELLNER. 1972. Vergleiche zwichen Hängebrücken und Schrägkabelbrücken für Spannweiten über 600 m. International Association for Bridge and Structural Engineering. Publication 32-I.
6. PODOLNY, W., JR. 1974. Cable Connections in Stayed Girder Bridges. Engineering Journal. American Institute of Steel Construction. Fourth Quarter. 11 (4): 99–111.
7. KONDO, K., S. KOMATSU, H. INOUE & A. MATSUKAWA. 1972. Design and Construction of Toyosatta-Ohhashi Bridge. Der Stahlbau. 41 (6): 181–189.
8. TAYLOR, P. R. 1969. Cable-Stayed Bridges and their Potential in Canada. Engineering Journal (Canada). 52 (11): 15–21.

CABLE-STAYED STEEL BRIDGES

Gerard F. Fox

Introduction

German engineers in the period following the end of World War II had the monumental task of rebuilding the long-span bridges destroyed during that war. However, they also had the opportunity to not only design modern new bridges but to provide for much more vehicle capacity. The destroyed bridges were for the most part narrow and would have been woefully inadequate to provide for the tremendous growth that occurred in post-war automobile and truck traffic.

Rather than perpetuating the bridge types of the past, the German engineers began their task afresh with an innovative spirit. The shortage of material and the need to use existing substructures led to the seeking of lighter and different type designs. The largest single technological advance made was the development of the orthotropic steel plate deck system.[1] This efficient floor system is utilized for most of the long-span steel bridges designed today.

Another innovation was the resurrection and refinement of the cable-stayed bridge type. The first known bridge of this type with all inclined stays was built 200 years ago of timber. Several more of metal construction were built soon after. Unfortunately, two of these bridges collapsed with the result that the public became disenchanted with this kind of bridge and the engineers turned to designing suspension bridges.

During the past 25 years there have been a large number (on the order of 70) of cable-stayed bridges constructed and at the present time a great number are in the planning and design stage. Today this type of bridge is considered competitive in the span length range of 600 to 2000 feet. They have practically eliminated consideration of truss bridges. This has occurred not only because of aesthetic reasons, since cable-stayed bridges are very striking in appearance (Fig. 1), but from economic considerations as well.

This paper contains a state-of-art review of the analysis, design and construction of steel cable-stayed bridges.

Types of Cable-stayed Bridges

The cable-stayed bridge is simple in its structural concept: essentially, it consists of a girder or truss bridge that is supported along its length at intervals between piers by steel cables which hang from high towers (Fig. 2). It differs from a suspension bridge in that the cables are not continuous from one end of the bridge to the other and the horizontal forces from the cable are for the most part balanced on each side of the tower, which negates the need of expensive anchorages as required for most suspension bridges. While there are suspension bridges that are self-anchored, they are difficult, and therefore

Gerard F. Fox is a Partner in the firm of Howard Needles Tammen & Bergendoff, New York, New York 10105.

0077-8923/80/0352-0087 $01.75/1 © 1980, NYAS

expensive, to construct since the deck system which transmits the compressive force must be constructed before the main cables are placed.

The cable-stayed bridge can consist of three or more spans with two or more towers (FIG. 2) or 2 spans with one tower (FIG. 3). There can be supporting or hold-down piers in the side span which increases the stiffness of the bridge (FIG. 4) or there may be a clear opening for the side span (FIG. 2). The deck is usually supported at the towers by bearing shoes but could also be completely free depending only on the cables for vertical support. The towers are usually fixed at their base but can also be pinned.

There are three main arrangements of cables, namely the radiating system in which the cables emanate from one point on the tower (FIG. 5), the harp

FIGURE 1. Proposed Dame Point Bridge, Jacksonville, Florida.

system in which the cables are parallel (FIG. 4), and the fan system which is similar to the radiating system except that the point support on the tower is expanded to a finite length of the tower height (FIG. 3).

There may be one plane of cables with the cables entering the deck span at the median between roadways (FIG. 3) or two planes of cables which may be vertical (FIG. 2) or sloping with attachments at the edges of the bridge deck (FIG. 5).

Alternate tower types may consist of simple double columns (FIG. 2) or single columns (FIG. 3), A-Frame (FIG. 5), Frame (FIG. 1) and modified A-Frame (FIG. 6). These towers can be constructed of steel plates or concrete depending on their comparative cost.

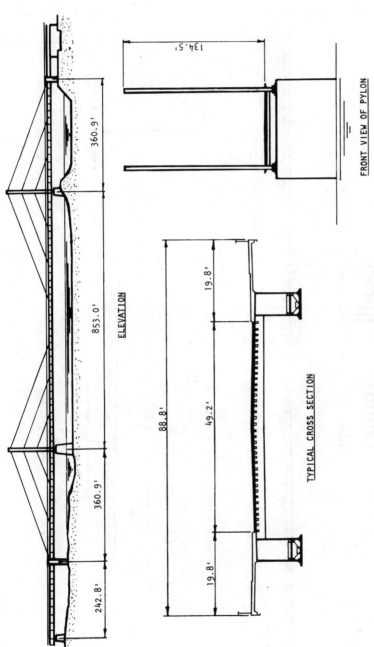

FIGURE 2. Theodor-Heuss Bridge at Düsseldorf.

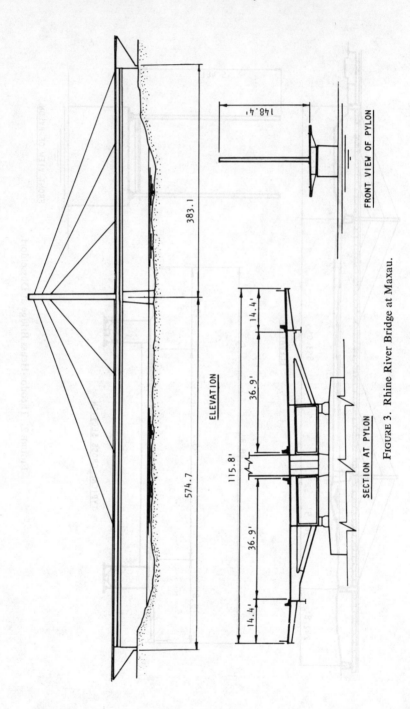

FIGURE 3. Rhine River Bridge at Maxau.

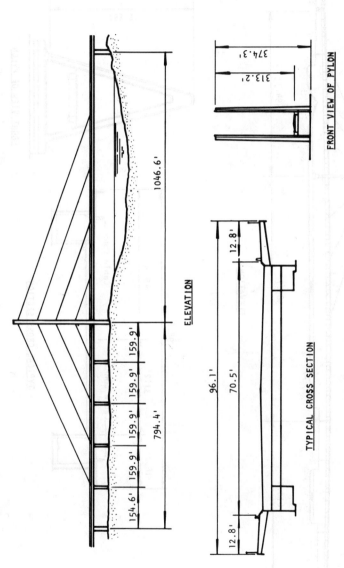

ELEVATION

TYPICAL CROSS SECTION

FRONT VIEW OF PYLON

FIGURE 4. Kniebrücke-Düsseldorf.

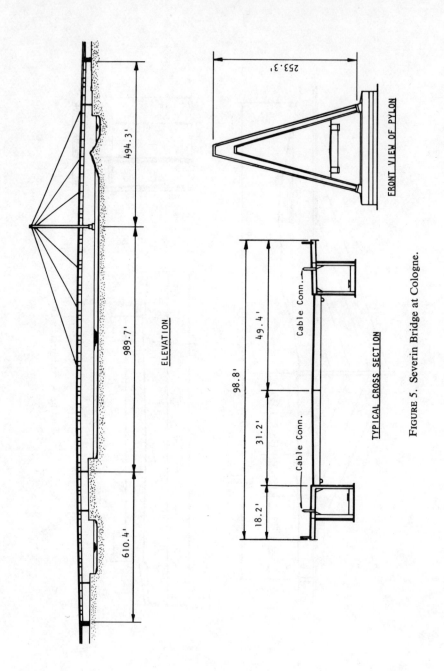

FIGURE 5. Severin Bridge at Cologne.

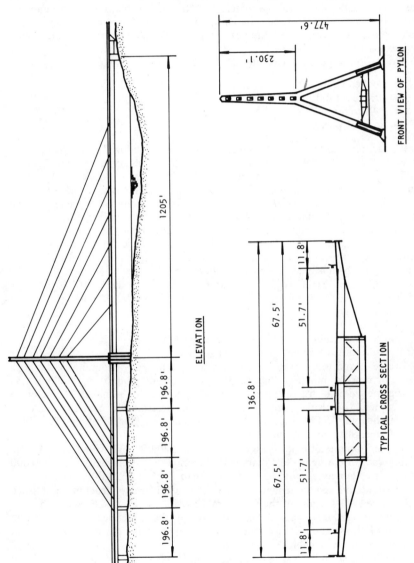

ELEVATION

FRONT VIEW OF PYLON

TYPICAL CROSS SECTION

FIGURE 6. Rhine River Bridge at Düsseldorf-Flehe.

CABLES

The selection of the cable system is probably the most important decision that must be made by the designer. A complex array of interrelated factors must be evaluated including the material of the main cable, the cable's anchorage and system of protection. In addition, schemes for the erection of the cables must be studied and an economic comparison of the various cable systems made.

The various types of steel cables that have been used include:

- Locked Coil
- Bridge Strand
- 7-Wire Strand
- Parallel Wires

Most existing European cable-stayed bridges have utilized preformed, prestressed, lock-coil strand as the cable elements. This type of cable is not readily available in the United States, and for this reason probably will not be considered for use on future cable-stayed bridges in the United States.

In the locked-coil strand, the individual wires of the outer layers of the cable are shaped to interlock and form perfectly solid annular rings. In the longitudinal direction the wires are spirally wound about the center wire.

As load is applied to the locked-coil strand, transverse forces are developed between the individual wires, which tend to make a tight fit at the exterior face of the cable, which is meant to be watertight.

The German Steel Highway Bridge Specification DIN 1073 indicates 228,000 psi as the minimum ultimate strength of the cable. For design, a value is specified of:

   .42 of ultimate strength for dead and live loads
   .46 of ultimate strength for other combinations
The modulus of elasticity is 24,200,000 psi.

Bridge strand is an assembly of wires formed helically around a center wire in one or more symmetrical layers. It is also spirally wound longitudinally, as are the locked-coil strands. Bridge strand is available from $\frac{1}{2}''$ diameter to $4''$ diameter. It is usually furnished galvanized, prestretched, and end-socketed. The applicable ASTM spec is A586–68—Standard Specifications for Zinc-Coated Steel Structural Strand.

The minimum ultimate tensile strength is 220,000 psi with Class A zinc coating and for diameters less than $2\frac{5}{8}''$, the modulus of elasticity is 24,000,000 psi and for diameters equal to or greater than $2\frac{5}{8}''$ it is 23,000,000 psi.

The tentative recommendation for cable-stayed bridge structures published in the May 1977 Journal of the Structural Division of the ASCE, states that the allowable design load should be one-third of the effective design breaking strength of the strand. This factor of safety of three is high compared to that used for parallel wire cables.[2]

The 7-wire strand is a stress-relieved strand and is the same as that furnished for prestressed concrete. It is available in two grades, Grade 250 and Grade 270, having minimum ultimate strength of 250 ksi and 270 ksi, respectively. The diameter of each strand ranges from .25″ to .6″.

Suspension bridges in the past have had their parallel wire strands spun in the field. This was an expensive and time-consuming operation. Now an alternative is available in the form of shop-assembled strands of parallel wires. These prefabricated strands are also used for cable-stayed bridges. The suspension bridge wire specified has a minimum ultimate strength of 225,000 psi with a yield strength of 160,000 psi. The usual diameter of the wire is .196" and the working stress is from 38% to 40% of the ultimate strength, as noted in the tentative recommendation referred to earlier. A number of strands can be used to make one individual cable. The modulus of elasticity is taken to be 27,500,000 psi.

CABLE CORROSION PROTECTION

The type of corrosion protection provided for the steel cables has been strongly influenced by the make-up of the cable used. The main types of corrosion protection that have been used are as follows:

1. *Paint.* The use of paint alone for protection has been limited to cables consisting of locked-coil strands. There have been at least two reported incidences of problems with painted locked-coil strands. The cables of the Köhlbrand Bridge in Hamburg corroded because salt-laden water penetrated the cable. The cables are being replaced on this bridge, which was opened to traffic in 1974. The cables of the Lake Maracaibo Bridge in Venezuela have also corroded after 18 years of service.

2. *Galvanizing.* The use of zinc coating of cables has been very popular in providing corrosion protection for the individual wires of bridge strands and prefabricated parallel wire strands. Stika Harbor Bridge in Alaska utilizes galvanized bridge strands as the cables. In addition, the use of galvanized bridge strands is common English practice, as exemplified by the Wye and Erskine bridges. If properly maintained, this type of protection should last a long time.

3. *Polyethylene Pipe.* This type of pipe has been used to encase 7-wire and parallel wire strands. Strands and pipe are assembled together in the shop and then wound on wheels for shipment to the field. After the cable is in place, grout under pressure fills the area between the wires and inside wall of the polyethylene pipe.

Another application of polyethylene was utilized on the Papineau Bridge constructed in 1969 and the Hawkshaw Bridge constructed in 1968 (both are Canadian bridges). Each cable of the Papineau Bridge consists of 12 bridge strands. Each bridge strand has a continuous seamless covering of black polyethylene 0.21" thick. The polyethylene material meets ASTM Class C Type 1 Grade Material Specification and contains enough anti-oxidant to combat ozone and ultraviolet attack. The jacket was applied to the cables before socketing and then local areas were removed at sockets and saddles. Local patching of the cover was done by hot air stick-welding with polyethylene rod. This same type of protection is proposed for the Huntington, West Virginia, cable-stayed bridge.

4. *Elasto-Wrap.* The elasto-wrap system was initially developed by the American Bridge Company for use on suspension bridge cables. The steps in applying the covering are as follows:

- First, a coat of liquid neoprene is applied to the galvanized wires of the cable.
- Second, a double thickness of uncured neoprene is spirally applied with a 50% lap.
- Then, two coats of Hypalon paint are applied to the sheet neoprene.

This system will provide excellent protection against atmospheric attacks and is easily reparable. It is expected that the wrapping will not need repainting for at least 15 years. This type of protection has been proposed for a bridge in Malaysia.

5. *Acrylic Resin.* At the same time, the Bethlehem Steel Company was developing a plastic cable covering for suspension bridges, which was first used on the Bidwell Bar Suspension Bridge in California in 1965. The cables of this bridge were made up of 37 bridge strands. Steps in the process of applying the protective covering were as follows:

- Polyethylene cable fillers were used to round out the cable.
- A covering of nylon film was applied.
- One layer of glass-fiber mat and two layers of glass cloth tape in combination with several coats of a lucite acrylic resin were applied.
- A weather coat of lucite and a finish coat of lucite were applied. This type of protection was used on at least two Japanese cable-stayed bridges to protect prefabricated parallel wire strand cables.

What is needed is a cable system consisting of parallel prefabricated galvanized wires, Hi-Am anchorages, and a corrosion-resistant wrapping that can be applied in the plant. Perhaps marine technology will be of assistance in finding other alternatives. A possible alternative would be the use of epoxy-coated wires without a wrapping, but research is needed to justify such usage.

CABLE FITTINGS

End fittings for locked-coil strand, structural strands and prefabricated parallel-wire strands usually consist of a poured-zinc type of socket. The ends of the strand are "broomed out," cleaned, and then inserted in the "basket" of the socket so that every individual wire is surrounded by the poured molten zinc.

Recognizing that the fatigue strength of zinc-poured sockets was not too high, because of damage caused by the high temperatures, the BBR Co. of Zurich developed a new anchorage which is called the Hi-Am Anchorage. It has a much better fatigue resistance than zinc sockets and contains a special mixture of metal balls and epoxy resin. This type of anchorage can support very large loads. For example, 300—¼" diameter wires could support a design load of 1,600,000 lbs.

Cables consisting of multiple units of 7-wire strands rely on steel wedge anchors (of the Freysonnet type) for each individual strand. All strands of the cable are stressed simultaneously. This type of strand and anchorage is from prestressed concrete technology.

Cables are considered pinned at their ends for analysis purposes, while in reality they are fixed. Therefore any change in slope that occurs in the edge girder or tower due to deflection will create bending stresses in the cable. For a cable fixed at the ends and subjected to a change in angle of B the stress induced is as follows [3]:

$$s = 2B\sqrt{\frac{ET}{A}}$$

where:

      s :Bending stress induced in cable
      E :Modulus of Elasticity
      T :Axial tension force in cable
      A :Area of cable

The bending stress can be reduced considerably if a neoprene collar is used as an elastic support at the exit of the cable from the edge girder or tower.

STRUCTURAL ANALYSIS

The structural analysis of the cable-stayed bridge system is not too difficult, even though the structure is highly redundant. For Dead Load (the weight of the structure) tensions in the cables can be assumed and a static analysis made or a linear elastic model can be adopted and a computer program, such as STRUDL, employed to obtain the bending moments, shears, thrusts, and cable tensions.

For Live Load, the same model can be used to obtain influence lines and thereby the internal forces. For this analysis, the cables are considered as straight members along their chords. To compensate for the actual sag, an effective modulus of elasticity, developed by Ernst, is used.[4] To maximize the effective modulus, the weight of the cable should be kept to a minimum and the tension in the cable at a maximum.

In addition to the cables, the towers and deck are subject to nonlinear effects resulting from their deflections. While the resulting additional stresses are usually not significant, they should be calculated. This can be accomplished by using an incremental procedure in which a linear stiffness analysis is made and the deflections calculated. For the deflected position an equilibrium check is made at each node and the unbalanced force is applied as a fictitious force and the procedure repeated until equilibrium is essentially reached. The stiffness matrix for each step is formed taking into account the deflected shape. The final deflections and internal forces are the sum of the individual ones from each step.

AERODYNAMIC CONSIDERATIONS

The aerodynamic effects of wind must be taken into account for any long-span steel cable-stayed bridge. Wind tunnel testing is desirable since there is no reliable method at present to predict the response of this type of bridge to wind.

The static effect of wind is merely the pressure that the wind makes on the structure, while the dynamic effect results from the separation and turbulence of the wind as it passes by the structure.

There are two types of dynamic phenomena to be concerned with, vortex shedding and flutter. Vortex shedding is the formation of vortices at regular intervals in the wake of the bridge. They form alternately at the top and bottom of the bridge and give rise to fluctuating forces in the vertical direction. If the frequency of these forces is the same as the natural frequency of the bridge, then resonance occurs. The maximum amplitude reached depends on the amount of damping present in the bridge. They can cause discomfort to the users of the bridge, but in general their effects are not serious. Buckland and Wardlaw [5] have recommended for consideration the following limits of acceleration for bridge movement:

wind speed   0 to 30 mph—2% of g
             30 to 70 mph—5% of g
             over 70 mph—disregard.

Flutter is the aerodynamic coupling of vertical and torsional motion. It is characterized by large increases in amplitude with no or small increase in wind speed. Sectional models are usually used to determine the wind speed at which flutter occurs. For the proposed Dame Point Bridge (FIG. 1), a 1 to 60 scale sectional model, 7 feet long, was tested in the wind tunnel at the Low Speed Aerodynamics Laboratory, National Research Council (NRC), Ottawa, Canada. This program was accomplished under the direction of Dr. R. L. Wardlaw and Dr. H. P. A. H. Irwin of the NRC.[6]

The original cross section of the steel bridge exhibited marginal flutter behavior and unacceptable vortex shedding excitation.[7] The cross section was modified by adding vertical baffles at the quarter-point locations of the cross section, replacing the solid traffic parapet by an open-type railing, and providing for sharper leading edges on the fairings of the edge girder. The critical flutter wind velocities for various angles of attack are shown on FIGURE 7.

There is very little in the literature on dynamic response of cables to wind-induced forces that can be immediately applied or used with a great deal of confidence. There have been no reported wind-induced vibration problems of cables equipped with a neoprene collar at the exit of the anchorage. Two concrete cable-stayed bridges that had problems did not have the neoprene collar. The cables of the Brotonne Bridge in France vibrated from quartering winds with amplitudes of several feet. To remedy the situation, large dampers were mounted on frames at the deck level (FIG. 8) and connected to the cable.

REPRESENTATIVE EXAMPLES

Severin Bridge

The Severin Bridge (FIGS. 5, 9, 10), constructed in 1959, spans the Rhine River at Cologne. The reason for using only one tower was to complement the tall spires of the Cologne Cathedral which is on the left bank. It is probably the most famous cable-stayed bridge in the world.

The bridge carries four lanes of vehicular traffic, two streetcar tracks, and a bicycle path on each side of the roadway.

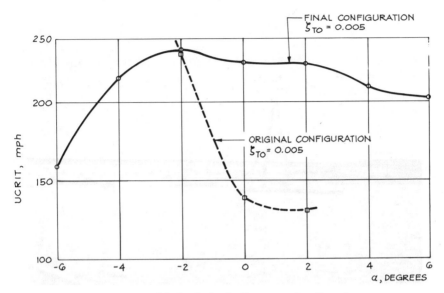

FIGURE 7. Critical velocity for flutter as a function of a.

FIGURE 8. Brotonne Bridge—View of shock absorber clamping device.

FIGURE 9. Severin Bridge at Cologne.

FIGURE 10. Tower of Severin Bridge at Cologne.

The steel A-frame tower is most striking in appearance with all of the locked-coil steel cables converging toward its top. The two main longitudinal box girders are only 15 feet deep at the center of the main span. The total weight of steel is 88 lbs per square foot, for a bridge length of 2266 feet and width of 94.8 feet, with 8.3% in the towers and 7.0% in the cables.[8]

Karlsruhe-Maxau Bridge

This cable-stayed structure (FIG. 3) that spans the Rhine River about 90 miles south of Frankfurt was constructed in 1966. It was the first bridge to have a single plane of locked-coil steel cables with a single pylon.

The bridge carries four traffic lanes with full-width parking shoulders on the right and a bicycle pathway on each side of the structure.

The stiffening girder has an average depth of 9.8 feet and is a single-cell steel box with cantilever wings supported by diagonal struts. There are also longitudinal girders at the edge of roadway to provide for load distribution.

The total weight of steel is 92 lbs per square foot for a bridge length of 958 feet and width of 114.1 feet, with 8.3% in the cables.[8]

Kniebrücke-Düsseldorf

The Kniebrücke (FIG. 4) was constructed in 1969 and is asymmetric with a double plane of locked-coil steel cables in a harp configuration. The pylons forming the pier rise to 313 feet above the roadway. The large span, 1047 feet, was erected by the free cantilever method.

The bridge has no median and carries three lanes of traffic in each direction. There are piers at each cable connection point in the side span to increase the stiffness of the bridge.

The two longitudinal plate girders are 9.9 feet deep and the floorbeams are 7.6 feet on centers.

The total weight of steel is 118 lbs per square foot for a bridge length of 1843 feet and width of 94.9 feet, with 19.4% in the towers and 11.6% in the cables.[8]

REFERENCES

1. CUSENS, A. R. & R. P. PAMA. 1975. Bridge Deck Analysis. John Wiley & Sons. New York. p. 29.
2. BIRDSALL, B. 1972. Discussion of paper, Mechanical Properties of Structural Cables, Journal of the Structural Division, ASCE, August. p. 1883.
3. WYATT, T. A. 1960. Secondary Stresses in Parallel Wire Suspension Cables. Journal of the Structural Division, ASCE, July.
4. PODOLNY, W., JR. & J. B. SCALZI. 1976. Construction and Design of Cable-Stayed Bridges. John Wiley & Sons. New York. pp. 350–352.
5. BUCKLAND, P. G. & R. L. WARDLAW. 1972. Some Aerodynamic Considerations in Bridge Design. Engineering Journal, April. p. 10.
6. IRWIN, H. P. A. H., M. G. SAVAGE & R. L. WARDLAW. 1978. A Wind Tunnel

Investigation of a Steel Design for the St. Johns River Bridge, Jacksonville, Florida. National Research Council, N.A.E., Report LTR-LA-220, February.
7. Fox, G. F. 1979. The Dame Point Bridge—Main Spans—Superstructure. Preprint, Boston ASCE Convention, April.
8. Weitz, F. R. 1975. Entwurfsgrundlagen und Entscheidungskriterien für Konstruktionssysteme im Großbrückenbau unter besonderer Berücksichtigung der Fertigung. Dissertation, Darmstadt.

STATE-OF-THE-ART
IN
CABLE-STAYED BRIDGES WITH
CONCRETE DECK STRUCTURE

T. Y. LIN AND CHARLES REDFIELD

CONCEPTUAL AND HISTORICAL DEVELOPMENT OF CABLE-STAYED CONCRETE BRIDGES

The four basic concepts that have led to the development of cable-stayed concrete bridges are shown in FIGURES 1 (a), (b), (c), and (d). These concepts are:

(a) The concept of a cantilever bridge that utilizes steel cable stays as supports for the superstructure, to replace intermediate piers.

(b) The concept of a cantilever bridge that utilizes steel cable stays in tension and the concrete deck in compression.

(c) A modification of post-tensioned double cantilever bridges made by relocating the slab tendons outside and upward, exposing them into external stays supported on towers.

(d) A modification of self-anchored suspension bridges by using straight stays in place of curved suspension cables and suspenders thus leading to elimination of stiffening trusses.

A list of cable-stayed concrete bridges appears in a paper by the ASCE Committee on that subject.[1] Referring to this document and others reveals that the development of long-span cable-stayed concrete bridges is relatively recent and there are far fewer of these bridges than of those with steel decks. There are several reasons for this late development. The relative weight of concrete has appeared to be a disadvantage in longer spans. Engineers with experience in long-span bridges have traditionally designed in steel and are not familiar with concrete both in design details and in erection of stayed structures. Furthermore, the weakness of concrete in tension was not overcome until prestressing became an accepted practice. As a result, although several short-span cable-stayed concrete bridges were built in the 1960s, the pioneering work in long-span cable-stayed concrete bridges was essentially done by Professor Riccardo Morandi alone. Thus, the first such bridges were the Lake Maracaibo Bridge completed in 1962 in Venezuela with five 215-meter spans (FIGURE 2 [2]) and later, the Wadi-El-Kuf Bridge in Libya (FIGURE 3 [3]).

A similar type was designed for the Mesopotamia Bridge in Argentina, 1972,[1] with a main span of 340 meters, but it has not yet been built. It is noted that these three bridges can be regarded as falling within concept (a) of FIGURE 1—the replacement of intermediate piers by large cable stays.

In order to lighten the deck concrete, the panel length between the stays was shortened and the number of stays per cantilever was increased. This

T. Y. Lin and Charles Redfield are with T. Y. Lin International, Consulting Engineers, 315 Bay Street, San Francisco, California 94133. T. Y. Lin is Chairman of the Board.

103

0077–8923/80/0352–0103 $01.75/1 © 1980, NYAS

happened in the case of the Rio Parana Bridge built in 1973 in Argentina with
a main span of 245 meters [4] and also in the Waal Bridge in Holland (FIGURE 4).
In the case of the Kwang Fu Bridge, designed by the authors (FIGURE 5) the
panels were made of precast I-beams supported on temporary bents, which also
served for the construction of the cross-girders. The Ruck-A-Chucky Bridge
shown in FIGURE 9 [7], which has single cantilevers starting from the hillside, is
a clear example of concept (b) of FIGURE 1.

Multiple-cable-stayed concrete bridges developed as a result of segmental

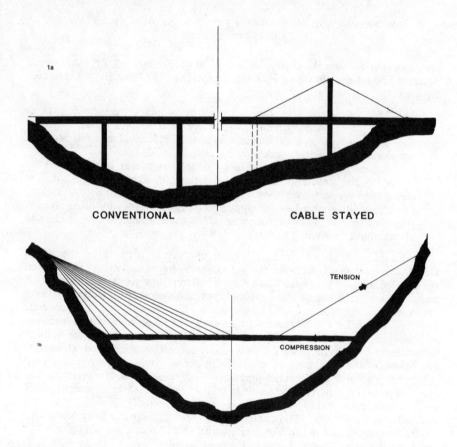

CONVENTIONAL   CABLE STAYED

TENSION

COMPRESSION

construction. The first major structure was the Brotonne Bridge in France
(FIGURE 6 [5]), with a span of 320 meters; it was immediately followed by the
Pasco-Kennewick Bridge shown in FIGURE 7 (Washington State, USA),[6] with
a main span of 299 meters. The proposed Dame Pt. Bridge, to be built in
Florida (FIGURE 8), has a main span of 396 meters. These can be considered
to be close to concept (c) of FIGURE 1, which extends the double cantilever
segmental construction to much longer spans. The above short history of
cable-stayed concrete bridges indicates that the optimum number of stays is
determined by the situation and each case should be considered by itself.

Number of Cable Stays

One advantage of the single-stay type is the necessity to construct only one set of anchorages and one transverse diaphragm for each cantilever. On the other hand, these anchorages become quite large, requiring massive transverse diaphragms. With multiple stays, relatively small anchorages similar to conventional post-tensioning tendons can be employed, utilizing lighter diaphragms distributed along the deck. When using multiple stays it is often possible to build the cantilever segmentally so that each segment can be immediately

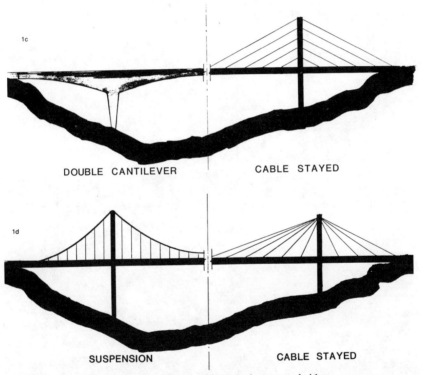

FIGURE 1. Concepts of cable-stayed concrete bridges.

connected to the post-tensioned cables. Thus, construction is carried on, either in the form of a traveling carriage for in-place concreting or using match-cast segments.

When using single stays, construction of the deck can proceed as follows:

1. The deck can be cantilevered as in conventional segmental double cantilever construction.
2. Precast girders can be erected between panel points which are temporarily supported on falsework until their incorporation into the final structural system (FIGURE 5).

3. Support trusses long enough to span between long panels can be constructed, as was done for the Waal Bridge near Thiel in Holland (FIGURE 4).

Another example of the single-stay layout is the Inter-Continental Peace Bridge (FIGURE 10),[8] designed by the authors. It will have 220 spans that are

FIGURE 2. Lake Maracaibo Bridge, Venezuela.

1,200 feet long each. To serve multiple transportation purposes, it will consist of large tubular sections, which can easily span some 400 feet between stays; hence a single-stay in this case will be sufficient and efficient. The erection of this bridge will be accomplished by barging each 1200-foot double cantilever span as one piece, after it is assembled from segments precast in the yard.

FIGURE 3. Wadi-El-Kuf Bridge, Libya.

FIGURE 4. Waal Bridge, Holland.

FIGURE 5. Kwang Fu Bridge, Taiwan. (T. Y. Lin International)

FIGURE 6. Brotonne Bridge, France (Courtesy of Campenon Bernard Cetra, Paris)

CABLE LAYOUTS AND ARRANGEMENTS

The number of stays for each cantilever has been discussed in the previous section and we will now look at some of the other problems involved.

1. *Cable Formation.* This can be of the fan type, radiating type, or the parallel-harped type, as has been used for steel cable-stayed bridges. The radiating type may cause congestion at the anchorages. Hence the fan type is often advantageous because it allows space for anchoring multiple stays on top of the columns, and it will also supply maximum heights for the best inclination of these stay cables.

2. *Steel for Cables.* There have been different types of steel for the cables: Bridge strands, parallel wires, high-tensile bars or 7-wire strands. For

FIGURE 7. Pasco-Kennewick Bridge, Washington State. (Arvid Grant)

long-span concrete bridges, the requirement for prestressing would tend to favor 7-wire strands because of the anchorages already developed. To control the stresses and the strains and lengths of the stays, post-tensioning is recommended and is usually necessary, because it is desirable to use standard tensioning equipment. The working stress in the steel strands can be as high as permitted for post-tensioning operations, as long as cable stiffness and fatigue effects are provided for. Grouting within steel tube casing is one way to increase cable stiffness. Fatigue can be controlled by limiting the stress range and providing dampening at end anchorages. Parallel wires of ¼ in diameter, with their patented forms, have been used, notably for steel bridges, but can also be used for cable-stayed concrete bridges.

FIGURE 8. Dame Pt. Bridge, Florida (Courtesy of Howard Needles Tammen & Bergendoff)

FIGURE 9. Ruck-A-Chucky Bridge, USA. (T. Y. Lin International)

The third steel form is high-strength steel bars. These require couplers and are generally more expensive, but have been claimed to possess certain advantages, such as additional stiffness and higher fatigue resistance at the anchorages.

3. *Erection of Cables.* The erection of these cables can be accomplished in various ways. One could erect a scaffold in case of single stays. Where multiple stays are required, the use of a pilot cable or a moveable ladder can aid construction.

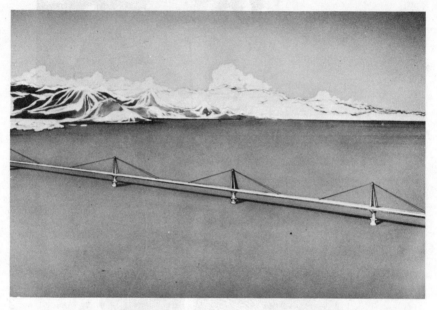

FIGURE 10. Inter-Continental Peace Bridge, Bering Strait. (T. Y. Lin International)

TOWER MATERIALS AND SHAPES

Bridge towers can be either steel or concrete, but it is generally agreed that particularly for cable-stayed concrete bridges, which are heavier, it is preferable to build the tower of concrete. The simple reason for this is that concrete carries compression more economically than steel. However, in order to take care of bending under various loading conditions, including wind and seismic stresses, the tower should be well reinforced. Concrete towers also provide more logical anchor detail for the stays.

The shape and the cross section of the towers is an individual problem and must be designed accordingly. For example, if the towers are made up of two columns placed outside the deck section, the size of the towers is not really limited. On the other hand, if a single column tower is designed, the space occupied by the column along the center strip of the bridge is restricted in its transverse dimensions. The buckling action of the tower above the deck area is an interesting problem. While the column seems like a cantilever supported at the bottom, it is actually elastically supported by the cables at the top. This elastic support not only restricts longitudinal movement of the column top, but also reduces the lateral buckling effect. This action has enabled designers to omit the expensive braces between the columns, as exemplified in the Double Canal Bridge (FIGURE 11).[9]. As mentioned in the previous section, the top of the tower should be designed to accommodate the cable stays, whether bent across the tower top in a saddle, or anchored thereto. The choice between the two arrangements will depend upon the length of the cable, the cable erection, and congestion at the tower top, as well as the desirability of stressing the stay cables from their lower ends below the deck. These are details to be worked out for each individual bridge.

DECK SECTIONS AND CONSTRUCTION METHODS

The design of the concrete deck section deserves a lot of study. There are several typical sections. For example, the closed rectangular or trapezoidal box section with internal struts or webs is a frequent one, because it affords excellent torsional as well as moment resistance in both the vertical and the horizontal directions.

The open section, such as used for the Pasco-Kennewick Bridge, has the advantage of being lighter. While it does not offer high moment resistance in the vertical plane and possesses small torsional resistance, such strength may not be limiting when employing multiple cable stays at close intervals, and when dynamic excitations under wind and seismic forces are not significant.

The use of interior precast struts to transfer local loadings to the cable stays was done for the Brotonne Bridge in Paris, and also planned for the Ruck-A-Chucky Bridge. This would tend to simplify forming and minimize the weight of the deck, which is a major problem for concrete bridges.

The method of construction is a prime consideration for cable-stayed concrete bridges. As previously mentioned, there are at least three choices:

1. One can pour the concrete in place on traveling carriages,[10] as is done for double cantilever segmental bridges. Both the carriage and the newly concreted deck will be supported by the stay cables as the cantilever progresses. When length of the panel is long between supports, the deck itself must be additionally prestressed and reinforced for erection.
2. Precast segmental pieces to be post-tensioned together can be used. The limit is the weight of the segments and the cost of the construction equipment.
3. The two methods can be combined, as was done for the Brotonne and the Ruck-A-Chucky Bridges mentioned above.

In all cases, the double cantilever construction, trying to maintain a balance between the two cantilevers, is considered an efficient procedure. Sometimes it is possible to build the anchor spans on falsework bents or to hold them into the mountainside by rock anchors, while the main span is to be cantilevered out.

The construction of the transverse diaphragm is a special consideration in itself. Frequently these diaphragms are heavily post-tensioned in the horizontal direction. While it is conceivable that the diaphragm could be built in structural

FIGURE 11. Strallato Bridge, near Vienna (L'Industria Italiana del Cemento, May 1979)

steel, the connection of the steel diaphragm to the concrete deck generally would mean additional cost and attention.

When using multiple cables, bending moments and shears in the concrete deck can be controlled so that they no longer present a major problem.

A good design approach is to balance the dead-load moments and shears, thus essentially leaving the deck concrete to perform only the job of absorbing the thrust created by the stay cables. Note that the axial compression force in the deck introduced by the stay cables is greatest near mid-span on account of the flat inclination of the cables for most types of arrangement. Therefore, it is important to reduce the weight of the concrete deck at mid-span. When we advance from midspan toward the piers, the thrusts accumulate from the cable stays and a larger section of concrete deck is needed to resist that axial compression. Hence it is necessary to gradually increase the concrete sectional area as one goes toward the piers. A variable sectional design is highly desirable because it lends itself to sectional area enlargement, particularly for long-span concrete bridges. In fact, the maximum span length of concrete deck construction is controlled by the ability of the deck to absorb the axial compression, more than anything else.

The live-load bending moment in the concrete deck is a relatively small item because the deck actually is supported elastically by the multiple stays. Of course, in case of single or double-stay layout, the bending moment would be more significant. For multiple-cable-stayed bridges, the critical vertical bending moment is at the point of rigid support furnished at the piers. If the deck is made continuous with the tower then heavy moments would be created at the junction. In the case of the Ruck-A-Chucky Bridge, the largest moment occurs at the abutment where the deck is fixed, as a result of its rigidity with the horizontal abutment. For the Pasco-Kennewick Bridge the deck was isolated from the piers in order to minimize its vertical moment. Live-load moments are not often a restricting factor in the overall deck design of cable-stayed concrete bridges. Even the theoretical live load, which is seldom realized in long-span bridges, will produce small global stresses in the deck, say on the order of 15% of the dead-load stresses. However, local stresses in the deck framing due to live load will be significant and must be analyzed.

Depending upon the location, wind and seismic excitations can be extremely important and should be investigated for each case. Of course, stresses produced by temperature changes, variations and gradient, as well as those due to shrinkage and creep, should be analyzed with respect to the construction sequence and local conditions. Depending upon the articulation employed, such stresses may or may not be controlling.

ECONOMICAL SPAN LENGTH FOR CABLE-STAYED CONCRETE BRIDGES

The economical span length for these bridges will be discussed in three parts, as follows:

1. *Cable-stayed Concrete vs. Cable-stayed Steel Bridges.* It must be first made clear that cost comparison between bridge types can only be presented in a general manner because conditions will vary for each bridge at each locality. Nevertheless, this discussion will bring out certain salient points that control the economics of bridge types for certain span lengths. To simplify

the discussion, the costs will be presented in dollars per square foot of deck surface area as they vary with the length of the main span.

All other things being equal, it is well known that the substructure cost per square foot of bridge deck actually decreases as the span increases. This is because certain constructional operations, such as cofferdams, pumping, etc. must be accomplished for each pier largely independent of the load to be carried by the foundation. Only when piles are required, or when the supporting stratum is relatively weak, will the substructure cost remain constant with varying span length, if expressed in terms of cost per square foot of deck. This is true for both the steel deck and the concrete deck, except that the steel deck will be much lighter and, therefore, will influence the substructure cost less.

The cost of the towers for a steel deck will be less than that for concrete deck. However, it is well known that a concrete tower will cost less than an equivalent steel tower. Therefore, unless concrete towers are also used for the steel bridge, the cost of towers may not be too different between a steel and a concrete bridge.

The cost of the cables is of paramount importance, particularly for long spans. If concrete deck is considered twice as heavy as the steel deck, the cost of these cables could double. Assuming that we maintain the same ratio of tower height to span length, the cost of cables will increase directly with the span length (in terms of dollars per square foot). This becomes a significant item when we deal with spans in excess of 1,000 feet and, in fact, it becomes almost a limiting item when spans exceed 2,000 feet.

When looking at the cost of the bridge deck itself, note that in order to span transversely and to transmit loads to the cable stays, the deck must possess a minimum section. This is true for both concrete and steel. That minimum section will usually suffice for spans up to about 1,000 feet, beyond which the concrete deck section will have to be gradually increased. The steel deck can remain almost constant perhaps up to about 1,200 feet beyond which the portion of the deck near the piers also require strengthening. However, the cost of the concrete deck is much lower than that of a steel deck. This cost differential offsets the savings in the cables, towers, and foundations resulting from the lighter steel deck.

The summation of these costs adds up to final costs, shown in FIGURE 12, for both the steel deck and the concrete deck, and a crossover point at about 1,500 feet is indicated. It must be emphasized that these costs are only approximate and will vary with many factors so that the numbers shown should not be taken for granted or used directly. This is particularly true since unit prices vary with times and locality and the cost of a bridge changes with many conditions. Nevertheless, the crossover point probably ranges between 1,200 and 2,000 feet.

It is interesting to note that in spite of the economic competitiveness of the concrete versus the steel deck, up to now most cable-stayed bridges have been built with steel and not concrete. Some engineers are unduly concerned with the inability of concrete to take tension, and the fact that either proper reinforcing or prestressing is required, whereas steel, by its very nature, can take both tension and compression so that one can feel more comfortable and can build into it a little more feeling of security. We believed that as knowledge and experience in the design of cable-stayed concrete bridges increase, many more applications will be found. No attempt is made here to compare cable-stayed concrete with cantilever steel bridges. It is assumed that the relative economics of cable-stayed steel bridges versus cantilever steel bridges can be utilized for such a comparison if desired.

  2. *Cable-stayed Concrete Bridges vs. Double Cantilever Concrete Bridges.* Although this is a topic that has baffled engineers, it can be stated that cable-stayed bridges will take over from the cantilevered segmental bridges for span length exceeding some 500 to 700 feet. An analysis of the cost of these two bridge types is shown in FIGURE 12, based on assumed unit costs. In FIGURE 13, the cost of the cantilever bridges is divided into the cost of foundation, pier, deck tendons and reinforcing bars, with the cost of plant equipment distributed among these various factors. The total cost of this type of bridge is also shown and is compared with the cost of the cable-stayed concrete bridges obtained from FIGURE 12. These two curves happen to cross over at about the 600-foot point. However, the assumed cost data are not really applicable to all cases.

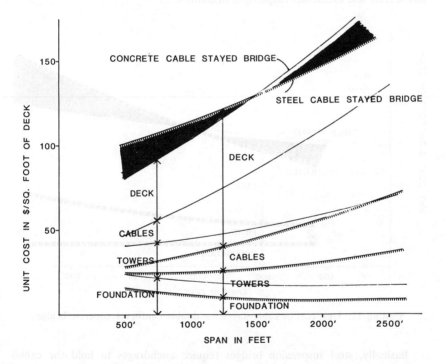

FIGURE 12. Unit cost of cable-stayed steel vs. concrete bridges.

To illustrate this point, cantilever concrete bridges have been built with up to about 800-foot spans, whereas cable-stayed concrete bridges have also been built for spans much shorter than 500 feet.

  In addition to the matter of cost, it is important to remember that cantilevered bridges are essentially open deck bridges, offering no structural obstruction above the deck. However, with long spans a box depth in excess of 30 or 40 feet will be needed at the support, whereas a multiple-cable-stayed bridge of even longer spans may only require a depth of deck not exceeding some 8 feet. On the other hand, cable-stayed bridges, being through bridges, produce obstruction above the deck that may hinder traffic movements, aviation, etc. Thus, in addition to the matter of construction cost, there are numerous other factors that may decide the type of bridge to be chosen for a given location.

3. *Cable-stayed Concrete vs. Steel Suspension Bridges.* Looking at recently built, medium-long-span bridges, one can conclude that cable-stayed steel bridges, and therefore cable-stayed concrete bridges (by inference from the previous discussion), would certainly be competitive with steel suspension bridges when spans of up to 1,500 feet are involved. For engineers familiar with the problem, one may judge the crossover point to be between 1,800 and 2,400 feet. When approaching 2,000 feet and beyond, the structure requires increasingly large concrete sections to absorb the compression near the piers. Constructional problems may become more complicated, giving way to cable-stayed steel and eventually suspension structures.

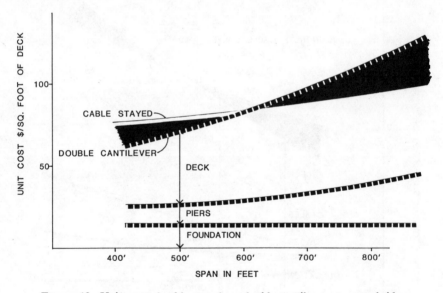

FIGURE 13. Unit cost of cable-stayed vs. double cantilever concrete bridges.

Basically, steel suspension bridges require anchorages to hold the cables against their horizontal pull, and sometimes abutments to deflect the horizontal component of the cables. These are not needed for cable-stayed bridges. Another costly item for suspension bridges is the steel deck system and its stiffening trusses.

It has been pointed out by the authors and other engineers that the cost of the suspension cables and suspenders will be almost double that of cable stays for cable-stayed bridges. One reason is that cables are of uniform section throughout the entire length of a suspension bridge and are controlled by the inclined tensile force in the cables at the tower top. For cable-stayed bridges, the amount of cables decreased as we get away from the towers. Additional length of cables is required for anchoring beyond the abutments of a suspension bridge. The erection of suspension bridge cables and deck requires a special skill not always readily available.

In conclusion, one can state that the era of cable-stayed concrete bridges is only beginning, although most of the techniques for design and construction have already been developed. The economic range for this type of bridge is probably between 500 feet and 2,000 feet. It is also very suitable economically for spans between 600 feet and 1,500 feet, particularly when a through-type structure is acceptable and thin deck depth is a requirement.

SUMMARY

This paper presents a brief survey of the development of cable-stayed concrete bridges. Design and construction considerations, including cable layout and arrangement, tower materials and shapes, deck sections and construction methods, and deck precasting as compared to in-place concreting have been discussed. Studies of the economics of cable-stayed concrete bridges relative to cable-stayed steel bridges, cantilever concrete bridges, and steel suspension bridges reveal the range of span and circumstances suitable for this type of bridge.

REFERENCE LIST

1. SUBCOMMITTEE ON CABLE-STAYED BRIDGES. Bibliography and Data on Cable-stayed Bridges. J. of the Structural Division, ASCE. October, 1977: 1971-2004.
2. SIMONS, HANNS, HEINZ WIND, W. HANS MOSER. 1966. Bridge over Lake Maracaibo, Venezuela. In The Concrete Architecture of Riccardo Morandi. 143-174.
3. DOMPIERI, I. 1973. Der Bau der Wadi-El-Kuf-Brucke in Libyen. Schweizerische Bauzeitung. Marz: 257-272.
4. "General Belgrano Bridge" over the Rio Parana, Argentina. VSL pamphlet. December, 1973. (Chaco-Corrientes Bridge).
5. MATHIVAT, J. 1978. The Brotonne Bridge. In FIP Eighth Congress 1978 Proceedings. 164-172.
6. GRANT A. 1979. The Pasco-Kennewick Bridge. J. PCI May/June: 90-111.
7. LIN, T. Y. & D. A. FIRMAGE. 1978. Design of the Ruck-A-Chucky Bridge. ASCE Reprint No. 3305. Oct. 1978. Chicago Convention.
8. T. Y. LIN INTERNATIONAL. Intercontinental Peace Bridge. Pamphlet.
9. PAUSERM, A. 1979. Ponte Strallato in c.a.p. sul canale del Danubio presso Vienna. L'Industria Italina del Cemento, May: 309-332.
10. West Germany: Prestressed stays plus three-phase concreting, gain time and money on unique bridge (Main River). Construction Management & Engineering, July 1973: 66-67.
11. PODOLNY, W. JR. & J. B. SCALZI. 1976. Construction and Design of Cable-stayed Bridges. John Wiley & Sons.
12. LEONHARDT, F. 1979. Cable-stayed Bridges. Deuscher Betontag Berlin, April 26. 1-11.

LONG-SPAN CONCRETE SEGMENTAL BRIDGES

Discussant: JEAN M. MULLER

The concrete segmental bridge is extremely economical and practical for long-span bridges. In our discussion we will review the many applications for concrete segmental bridges and show examples of many long-span bridges designed by the author. These bridges are in France, America and other parts of the world.

SCOPE: VARIOUS TYPES OF LONG-SPAN CONCRETE BRIDGES

The various types of bridges and their probable maximum span length are shown as indicated

	Types of Bridges	Probable Maximum Span Lengths
Figure 1.	Box Girders	1,000′
Figure 2.	Trusses	1,200′
Figure 3.	Frames (with Slant Legs)	1,200′
Figure 4.	Arches	1,500′–2,000′
Figure 5.	Stayed Bridges	1,500′–2,000′ (Depending on available construction methods)

All concrete bridges shown are built in cantilever with either precast or cast-in-place segments. The probable maximum span lengths are based on material available today.

GIRDER BRIDGES (FIGURE 1)

Jean Muller designed the first matched cast segmental bridge with epoxy joints—Choisy-le-Roi (over the Seine River), completed in 1962, as well as many cast-in-place or precast concrete segmental bridges that illustrate the many applications of the segmental bridge.
Some examples of precast segmental bridges designed by Jean Muller:
Linn Cove Viaduct—(America—Blue Ridge Parkway, North Carolina. Required to build bridge from the top. One directional cantilever/progressive placing/180-foot spans, horizontal and vertical curves, precast segmental piers)
Oleron Viaduct—(France—10,000 feet long, designed and built in 25 months, construction completed in May 1966, most of the spans are 270 feet, variable depth superstructure, first time launching girder was used to place segments from above.)
Chillon Viaduct—(Switzerland—320-foot spans, variable depth superstructure with horizontal and vertical curve, completed 1967. Jean Muller was consultant for design and construction)

Jean M. Muller is Chairman of the Board and Director of Research and Design for Figg and Muller Engineers, Inc. Tallahasee, Florida 32301.

0077-8923/80/0352-0123 $01.75/1 © 1980, NYAS

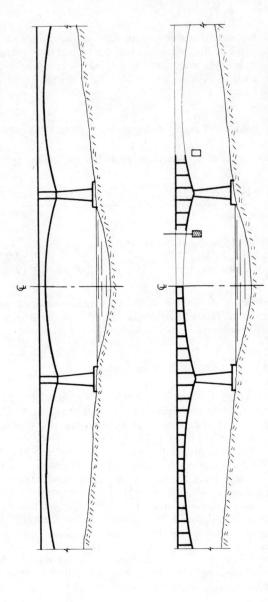

FIGURE 1. Long-span girder construction. Precast or cast-in-place cantilever construction. (Figg and Muller Engineers, Inc.)

St. Cloud Bridge—(France—337-foot main span over Seine River on 1,090-foot horizontal curve. Segments weight = 130 tons. Roadway width = 67 feet. Construction completed 1972)

St. Andre de Cubzac—(France—variable depth superstructure with 320-foot spans. Superstructure segments placed by beam and winch method)

Sallingsund Bridge—(Denmark—320-foot variable depth spans, 5,500 feet long, completed 1977)

Some cast-in-place bridges designed by Jean Muller:

Magnan Viaduct—(France—400-foot main spans with variable depth superstructure. Twin piers 350 feet high.)

Gennevilliers Bridge—(France—main spans are 600 feet long, crossing the Seine with variable depth superstructure and horizontal curve—completed in 1977.)

Houston Ship Channel—(America—750-foot main span, 375-foot side spans, variable depth superstructure. Figg and Muller Engineers, Inc., are consultants to the contractor in optimizing this design—construction started mid-1979.)

Proposed Scheme for Great Belt in Denmark (FIGURE 6) has 1,070-foot concrete segmental spans with a 980-foot side span.

TRUSSES (FIGURE 2)

The purpose and advantages of trusses are as follows:

- Reduce weight of web
- Simplify casting very high segments
- Reduce unit weight of precast segments

The main limitation is:

- Complication of connections between diagonals and chords.

This type of bridge was first attempted by Eugené Freyssinet in 1948. Jean M. Muller was the designer working directly with Freyssinet on the Harrach Bridge near Algiers, Algeria.

The Brisbane Bridge in Australia is an example of this construction.

The method certainly has application between the optimum span lengths of typical box girders on the low range and stayed bridges on the high range.

FRAMES WITH SLANT LEGS (FIGURE 3)

When the configuration of the ground allows, the use of inclined legs reduces the effective span length. Provisional back stays or a temporary pier are needed to permit cantilever construction. This requirement can sometimes become a difficulty.

The Bonhomme Bridge in Brittany is an example of this design.

Such a statical scheme is a transition between the box girder bridge with vertical piers and the true arch where the load is carried by the arch rib along the pressure line with minimum bending and the deck is supported by spandrel columns.

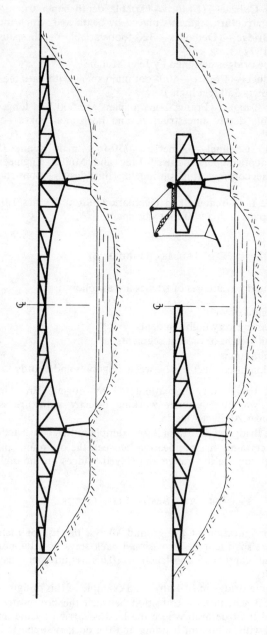

FIGURE 2. Long-span concrete trusses. Precast or cast-in-place cantilever construction. (Figg and Muller Engineers, Inc.)

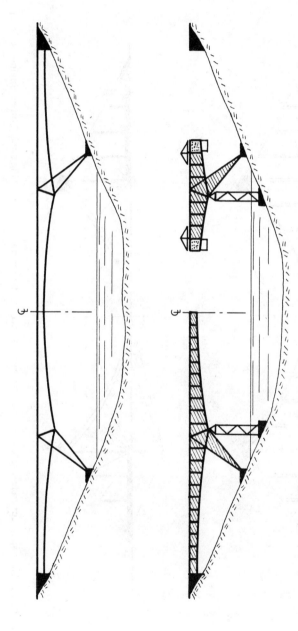

FIGURE 3.   Long-span concrete frames.   Precast or cast-in-place cantilever construction.   (Figg and Muller Engineers, Inc.)

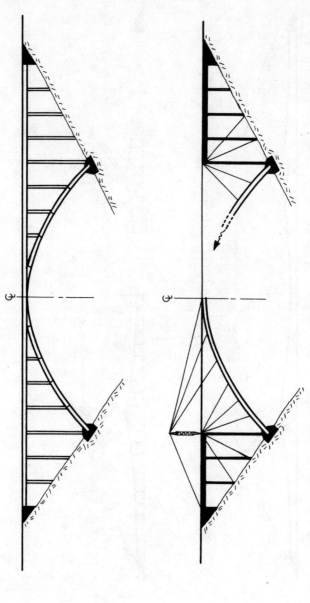

FIGURE 4. Long-span concrete arches. Precast or cast-in-place cantilever construction. (Figg and Muller Engineers, Inc.)

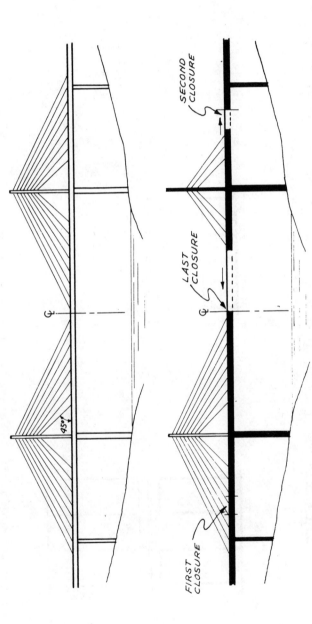

FIGURE 5. Long-span concrete-stayed bridges. Precast or cast-in-place cantilever construction. (Figg and Muller Engineers, Inc.)

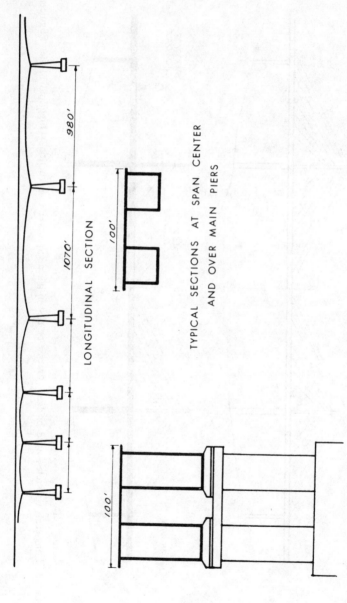

LONGITUDINAL SECTION

TYPICAL SECTIONS AT SPAN CENTER
AND OVER MAIN PIERS

FIGURE 6. Proposed concrete segmental bridge for the Great Belt project. (Figg and Muller Engineers, Inc.)

ARCHES (FIGURE 4)

Arches are the most economical scheme to carry loads to the ground where foundation conditions are adequate to resist the horizontal component of the reaction.

Eugené Freyssinet had prepared a design for a 1,000-meter span (3,300 feet) 40 years ago; but the maximum span built remained no more than 1,000 feet up to this year because of construction difficulties. Construction of a false-work is usually prohibitive. Progress was made only when cantilever endloads were applied to arch construction.

Various examples of arch construction are as follows:

- Caracas (500-foot span)
- Gladesville (1,000-foot span)
- Magnan (850 feet) *not built*
- Kerk in Yugoslavia, maximum span 390 m (1,280 feet—exactly the span as cable-stayed bridge—Dame Point, Jacksonville, Florida)

STAYED BRIDGES (FIGURE 5)

The concept of stayed bridges gained immediate acceptance when construction was made possible in balanced cantilevers with a large number of stays uniformly distributed along the deck.

The Brotonne Bridge is an excellent example of a stayed bridge. It is the longest concrete span built to date—1,050-foot center span. Jean M. Muller designed this stayed bridge with one line of stays down the centerline of the bridge. Construction was completed on this 4-lane bridge in 1977.

STATE OF THE ART AND FUTURE

1. Most of the new bridges in the USA have sites that call for a segmental concrete design.

2. On the basis concrete segmental bridges that have been built plus the design and construction experience that exists today, concrete spans of the order of 1,500 to 1,600 feet may be considered immediately feasible (see FIGURE 6).

3. Further development will move this limit upward.

4. The inherent advantages of concrete:

- Aerodynamic stability
- Aesthetics
- Low maintenance
- Lowest initial cost.

PERFORMANCE OF LONG-SPAN
BRIDGE STRUCTURES:
THE EAST RIVER BRIDGES AT NEW YORK

BLAIR BIRDSALL

My remarks will be limited to the four major East River crossings—the three grand old work-horses: Queensboro, Williamsburg, and Manhattan—and the city's crown jewel, the Brooklyn Bridge. The first three have served the community long and well through many changes in the modes of transportation for almost three-quarters of this century.

The Brooklyn Bridge has reigned supreme for almost a century. It will be 100 years old in 1983.

A short introduction to each will serve to give more life to the words.

The Queensboro Bridge (see FIGURE 1) crosses the East River on an alignment between 59th and 60th Streets, joining the Manhattan and Queens sections of New York City. Some of its piers are located on an island, which was known as Blackwell's Island when the bridge was built, was called Welfare Island for many decades, and has now been re-christened Roosevelt Island.

The Queensboro is a cantilever bridge consisting of five principal spans. Two are cantilever spans over the main channels of the order of 1,000 feet each. The two end spans provide the anchorage for one side of each cantilever span, and the center span provides the other anchor for both cantilever spans.

The Williamsburg Bridge (see FIGURE 2) is a suspension bridge connecting Delancey Street in Manhattan with the Williamsburg section of Brooklyn. It has a main span of 1,600 feet and straight backstays. The spans flanking the main span are supported from the ground, not from the cables. Thus the cables are relatively straight from tower tops to anchorages. It has four main cables.

The Manhattan Bridge (see FIGURE 3) connects Canal Street in Manhattan with Brooklyn. It has a main span of 1,470 feet and has suspended side spans. It also has four cables.

The first decade of the 20th century was a golden age for construction. These three bridges were all built during that decade, as was the city's first subway.

The Brooklyn Bridge (see FIGURE 4) was completed in 1883. It also is a three-span suspension bridge, with a main span of just under 1,600 feet, a truly remarkable feat at the time. It also has four cables.

The bridges have frequently been called upon to adjust to new conditions. When the Brooklyn Bridge was completed, motorized traffic was unknown, and rail traffic was very primitive.

The Queensboro Bridge has always had eleven travel ways usable for some sort of wheeled traffic. When it opened in 1909, it contained six rail lines, three vehicular lanes, and two pedestrian walks. Now all eleven lanes are used for motor vehicles.

Since its inauguration in 1903, the Williamsburg Bridge has also been

Blair Birdsall is a partner of Steinman, Boynton, Gronquist & Birdsall, New York, New York 10004.

0077–8923/80/0352–0133 $01.75/1 © 1980, NYAS

FIGURE 1. The Queensboro Bridge.

FIGURE 2. The Williamsburg Bridge.

FIGURE 3. The Manhattan Bridge.

FIGURE 4. The Brooklyn Bridge.

subjected to changes in traffic pattern. At present eight lanes are devoted to vehicular traffic and two lanes to rail traffic.

The Manhattan Bridge has always had eleven travel ways for wheeled traffic and two lanes for pedestrians. When it opened in 1909, there were eight rail lanes and three vehicular lanes, plus the two pedestrian lanes. Changes have been made over the years, and there are now seven lanes for motor vehicles and four for rail transit. The pedestrian lanes have been discontinued.

The Brooklyn Bridge (see FIGURE 5) has always had six travel ways for wheeled traffic, plus the great central promenade. On opening day in 1883, two ways were used for rail traffic and four for horse drawn vehicles and cattle. In later years, four of the six lanes were used for rail traffic. Since 1952, after a major re-arrangement of the roadway structure, all six lanes have been used for light motor vehicle traffic.

Thus, these great bridges have all performed well, adapting admirably to the changing fashions in the modes and character of transportation, and resisting the ravages of a savage environment. That they are still standing and still viable for the needs of the present day, is a great tribute to the foresight of their designers and the perseverance, dedication, and skill of those charged with their maintenance.

The latter has not been easy. These bridges, like thousands of others around the country, are all tax-supported, toll-free structures. For maintenance funds they have been competing with welfare, schools, hospitals, potholes, and solid waste collection, to say nothing of fire and police protection and the general running expenses of the city. These are powerful competition indeed.

Funds have not been available nationally for regular preventive maintenance programs. Maintenance forces have found it necessary to operate like fire departments.

At long last public awareness has been aroused to the plight of the nation's bridges. Funds have now been made available, and these bridges are finally receiving the attention they deserve—full in-depth inspection, rating, updating, and rehabilitation by qualified engineering firms. This paper consists of a progress report on those activities.

QUEENSBORO BRIDGE

The in-depth inspection has now been completed, assimilated, and has become the basis for recommendations for rehabilitation.

There is no indication of potential danger to primary supporting members, but much work needs to be done on certain details.

Items requiring immediate attention have been brought to the attention of the city, whose engineers are reacting immediately with proper safeguards.

For the remainder of the work, which can await normal design and contract procedures, the items have been given priorities and will be accomplished on a schedule, which has not as yet been determined.

The first items will be rehabilitation of the old off-ramp from the upper roadway to 21st Street in Queens, the so-called "Old Approach B," leading from Thompson Avenue, Long Island City to the Manhattan-bound upper lanes; and the two lower outer roadways, which are temporarily closed on the Main Bridge.

Many other lanes, ramps, and other details will be included among the

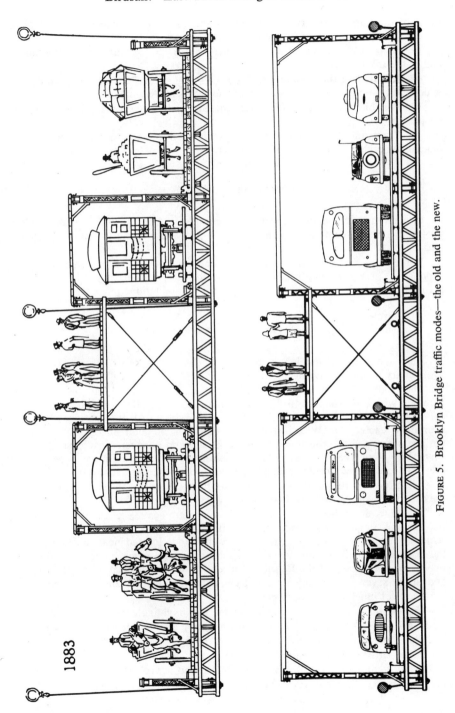

FIGURE 5. Brooklyn Bridge traffic modes—the old and the new.

items of decreasing priority. Observers of the New York scene will be interested to know that the following items are being considered:

1. Softening or removal of the unpleasant "S" turns near the Manhattan end of the Upper Roadways.

2. Removal of the unsightly and unused forest of steel beams and columns, which clutter up the Manhattan approach near Second Avenue.

3. Possible use of the lower outer lanes for pedestrians and cyclists during off-peak hours.

4. Changing the main lower roadway from five lanes to four lanes, replacing the middle lane with a safety barrier.

WILLIAMSBURG BRIDGE

Inspection is well under way, but not complete. Once again, primary supporting members seem to be without serious problems, to the extent inspected to date.

Immediate short-term measures were found necessary on the outer roadway supports. These are being taken care of by the city, pending normal development of plans and specifications for long-term rehabilitation.

MANHATTAN BRIDGE

Here again, inspection is still under way, but no signs of serious trouble for primary members have been uncovered.

This bridge, and the Brooklyn Bridge, require special structural studies. In common with the Brooklyn Bridge, this bridge has a complication that does not exist on the Williamsburg, the other suspension bridge. All three bridges have four main cables. On Williamsburg, by good luck or good management, the designers grouped the four cables into two pairs and connected each pair to a single line along the roadway. On Manhattan and Brooklyn Bridges, each of the four cables is connected along a separate line along the roadway.

This produces the need for a careful study to determine the distribution of the dead load, and the stresses caused by it.

The Manhattan Bridge has one additional complication peculiar to itself. By bad luck or poor management, the four subway tracks now in use on the structure are arranged in two pairs, and each pair is located near one edge of the bridge. If one is standing at one edge of the bridge at mid-main-span when two trains are passing at the far edge, the guardrail may be seen to move down and up by several feet. This does not constitute a threat to the bridge, but is the source of much maintenance cost.

The existence of these special problems was highlighted by an inspection made during the 1950s, and by a concurrent engineered construction job, which included replacement of cable bands and suspenders.

In the early 1970s, a study was made to find means of reducing torsional deformation due to the subway trains. The findings of that study will be reviewed and confirmed or revised in the current investigation.

BROOKLYN BRIDGE

Here also inspection is still under way. At the same time, however, it has been possible to advance the design of certain items, the need for which was apparent at the outset.

As to inspection, there is not as yet need for any concern regarding the primary supporting members.

The main cables are in excellent condition at the tower tops. Full inspection has not yet been made at critical points in the spans, but indications are favorable. There is considerable wire corrosion at the anchorages. The remaining strength is now being determined, but indications are that it is adequate to support the load over the long term, that it is reparable, and that it will be possible to arrest further progress of corrosion.

As to items in the design process, the following may be of interest:

Short Suspenders

At mid-main-span the short suspenders between cable and suspended structure consist of very short fixed-length rods.

The peculiar structural arrangement of the bridge requires that these suspenders rotate through a large angle because of changes in temperature. The original design did not provide adequately for this service, which would require frictionless rotation, and extensibility.

These features are being incorporated in design thinking in the event that replacement is required. At the same time, the structural system is being reviewed to determine the feasibility and desirability of a change, which would obviate the need for such special suspenders.

Diagonal Stays at Tower Top

The diagonal stays have been subjected to severe environmental attack at the point where they all join at the tower top. Design studies are being made as to a means of terminating all the strands at or near the edges of the tower, where they can be reached for observation and maintenance.

These are four outstanding examples of the many thousands of bridges in this country, which are overdue for in-depth attention. The situation emphasizes the need at all levels of government to set aside adequate funds for sound preventive maintenance programs.

REPLACEMENT OF THE UPPER DECK OF THE GEORGE WASHINGTON BRIDGE

R. M. MONTI, EUGENE FASULLO, AND DANIEL M. HAHN

This paper will focus on the performance of the original Upper Deck of the George Washington Bridge over its 46-year life. As a result of detailed investigations during the early 1970s, it was concluded that the economic life of the deck had been reached and that major repair or replacement was necessary. The reasons for this conclusion will be discussed in detail with regard to annual maintenance cost, total area of deck repaired, chloride content, concrete strength, and delamination at the level of the top steel.

In addition, the other phases of the study concerning the analysis of alternative rehabilitation and replacement schemes, the test program and the actual replacement of the Upper Deck will be briefly discussed.

INTRODUCTION

The George Washington Bridge, which spans the Hudson River between Washington Heights, Manhattan and Fort Lee, New Jersey is one of the most heavily traveled bridges in the world with a present annual volume of over 79,000,000 vehicles (including 7,400,000 trucks) utilizing the bridge's fourteen lanes.

When opened to traffic in October, 1931, the George Washington Bridge was Othmar H. Ammann's crowning achievement and fulfillment of a life-long dream, since it doubled the main span of any previously constructed suspension bridge—3,500 linear feet between centerlines of towers.

Only six lanes of the Upper Level were completed during the original construction, although the main load carrying elements of the bridge, including the cables and tower structures were designed for future expansion. The steel bulb tees and tie-rods, which form the main slab reinforcement, were installed for the two future center lanes at that time; however, concrete was not placed for these lanes until 1946 when the increased traffic volume necessitated completion.

In the late 1950s, construction of the six-lane Lower Level was undertaken and that level opened to traffic in 1962.

It is estimated that between 1931 and 1977, when replacement of the Upper Level deck slab began, over 1.6 billion vehicles had used the bridge, with approximately two-thirds of this total having used the original upper roadway.

During its forty-six year life, the original deck has withstood, not only this heavy traffic volume, but also, the deleterious effects of weather (freeze-thaw cycles), de-icing salts, and steel-studded snow tires, while giving satisfactory service to the traveling public. Deck deterioration was just becoming noticeable in the late 1950s, and the first major repairs were made to the deck slab in 1961. Since then, an estimated 48% of the upper deck slab has either been patched or completely replaced. The rate of deterioration has almost doubled

R. M. Monti, Eugene Fasullo and Daniel M. Hahn are with The Port Authority of New York and New Jersey, One World Trade Center, New York, New York 10048.

from 1970–75 relative to 1961–70, and by 1976, the annual repair cost was averaging approximately $500,000 with two patching contracts required per year.

Several years ago, a detailed investigation was initiated by Port Authority engineers to determine whether a complete restoration or replacement was warranted. This investigation resulted in an evaluation of sixteen schemes for restoration or replacement of the Upper Level deck slab and ultimately in the replacement of the slab with a prefabricated, prepaved orthotropic steel deck.

ORIGINAL DECK SLAB AND SUPPORTING FLOOR SYSTEM

The Upper Level deck is 4,660 feet between anchorages, and 90 feet wide between curb lines, with two 15-foot-wide sidewalks. A 2-foot-wide steel median barrier divides the deck into two 44-foot-wide, four-lane roadways. The total area of the Upper Level deck slab is 405,500 square feet exclusive of the steel median barrier and steel expansion finger joints located adjacent to the towers.

The floor system of the deck is constructed as follows: the main floor beams are transverse plate girders 10 feet deep composed of silicon steel flanges and carbon steel webs hung from the suspender cables. These floor beams support eight lines of roadway stringers and two fascia girders.

The roadway stringers are 60 foot long longitudinal plate girders varying in depth from 5′ 4″ to 5′ 8″, whose flanges are composed of nonweldable silicon steel while the web plates, stiffeners, and other details are made of carbon steel. These members are simply supported, one end being attached and the other free to move longitudinally; seated connections are provided on the main floor beams to receive the stringers. These stringers are also held together in pairs with cross-frames at the expansion ends and at the center of the panel (see FIGURE 1).

The roadway stringers support transverse secondary floor beams (CB16s)

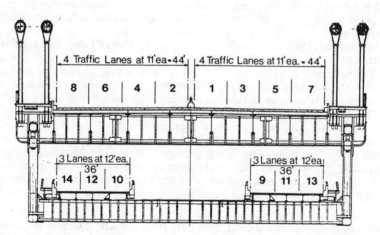

FIGURE 1. George Washington Bridge—Typical Cross Section (looking west).

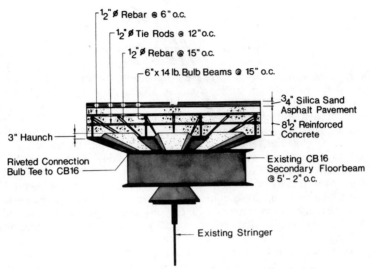

FIGURE 2. Typical section through existing deck slab.

spaced typically at 5'-2" o.c. The beams in the three middle bays are made of nonweldable silicon steel and those in the two outer bays are continuous over two spans and are made of structural carbon steel. All secondary floor beams are noncomposite and, therefore, were designed to resist both the dead and live loads acting independently of the concrete.

The original deck slab was an 8.5-inch-thick reinforced concrete slab with a ¾ inch thick silica sand asphalt pavement. The main reinforcement consisted of 6 in × 14 lb bulb tees at 15 inch o. c. which were continuous over the CB16s and riveted to the top flanges. The bulb tees were held transversely by ½" diameter tie rods. Top reinforcement consisted of #4s at 15 o. c. spaced between the bulb tees, and #4s at 6 in. o.c., the transverse reinforcement. The concrete was haunched up 3 inches between the bulb tees. The average weight of the slab, including pavement, was 106 psf (see FIGURE 2).

Although all slabs were cast-in-place, the six lanes placed in 1931 were formed with wood that was removed after construction, while the two center lanes installed in 1946 were cast into cinder concrete forms that were left in place and became part of the slab. The bulb tees were selected for the main reinforcement in the slab because they had a wider bottom flange for riveting to the beams below and because the small round top flange made it possible to work the concrete around them in a satisfactory manner and avoid a flat steel surface near the top of the slab.

The roadway concrete was originally specified to have a strength of 4,000 pounds per square inch after 28 days. All roadway slabs were designed to sustain a 25 ton truck with an 18,000 pound wheel load and 75% impact. This high impact factor was used because of the hard rubber tires used on vehicles of that era. This design loading has allowed the Port Authority to permit use of the structure for passage of a relatively high volume of overload vehicles during the life of the deck.

DETERIORATION OF THE UPPER DECK SLAB

Annual Maintenance and Repairs

Deck deterioration started to become noticeable on the upper deck of the George Washington Bridge during the late 1950s. It was not until after the severe winter of 1960–61 that a thorough inspection was made to determine the condition of the deck. The results indicated that deterioration was far beyond the scope of the bridge maintenance personnel such that contracts had to be awarded to expedite repair operations. During 1968, the deck was again repaired and resurfaced with a ¾ inch thick wearing course; despite these efforts, however, the deterioration of the deck continued to be a major problem.

Spalling and scaling of the concrete slab were the two types of deterioration common to the deck. Spalling, commonly known as a pothole, is the term indicating the separation and removal of large concrete fragments from the deck's surface. The top layer of deck reinforcement is often exposed. Scaling denotes a loss of surfacing, in many cases exposing the aggregate.

Repair practices included paving, patching and in some instances, full slab replacement. Paving the roadway serves little in the way of adding strength, and in some instances, is harmful in assessing the structural integrity of the deck. These repairs tend to cover any visible weaknesses such as potential potholes or cracks that normally could have been observed and repaired.

Although no records of previous patching repairs have been kept, it is suspected that some distressed areas have been subjected to repeated patching. There are a few possible reasons for this. The concrete patch, after it is poured, is subjected to the effects of shrinkage. This sets up an area of weakness on the boundary of the patch, that in time, can result in more spalling and another pothole. Another reason lies in the patching practice itself, when an area is prepared for patching, the depth of the slab is reduced. Since the slab is designed for a thickness of 5½ inches, a reduction of a few inches during repair will seriously affect it. The traffic in the adjacent lanes will exert forces on the reduced slab, thereby causing cracks. After the patch is set, these cracks propagate to the surface and another pothole develops.

It should be noted that in addition to the year-round repairs made by Port Authority's personnel, bi-annual deck repair contracts had been awarded between 1972 and 1975 at an annual cost by 1975 of over a half-million dollars. These annual contracts had attempted to prolong the life of the patch by specifying a 6 hour–2,000 pound concrete/7 day–4,000 pound concrete, with the perimeter of the patch treated with an epoxy bonding compound.

A review of the records on the amount of patching repairs performed on the Upper Level deck between 1961 and 1975 shows that an area equivalent to 48% of the total surface had been repaired. The amount of repair work performed between 1970–1975 (22½%) almost equals that performed over the prior 10 years (25½%) indicating that the rate of deterioration had almost doubled (see FIGURE 3).

Causes of Deterioration

A bridge deck must withstand the most damaging types of live-load forces, for example, the concentrated and direct pounding of truck wheels. A function of the deck is to distribute these forces in a favorable manner to the support elements below. The ratio of live to total load stresses is high in the bridge deck, usually much higher than in most of the other components of the bridge, and such fatigue-producing stresses tend to aggravate any defects that might be present in the deck. Because of its exposed location, temperature variations are larger in the deck and restraints to the resulting volume changes tend to cause cracking of the concrete. Salt has pronounced effects on the concrete, both for scaling and spalling.

It should be understood that salt is not the only cause of scaling and spalling. However, in contemporary bridge deck experience, the preponderant amount of deterioration is caused by de-icing salts. Investigation reveals corrosion of reinforcing steel by the de-icing salts is the primary cause of potholes. Once corrosion of steel occurs, the corrosion products can occupy 2.2 times as much

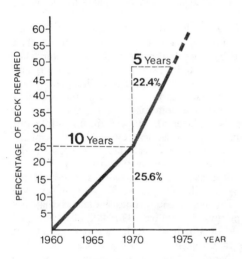

FIGURE 3. Year vs. cumulative % of deck repaired—1961 through 1975.

space as the original metal and may develop mechanical pressures as high as 4,700 psi, a force many times the tensile strength of concrete. The result of this pressure caused by rust is either the cracking and separation of concrete over the bar or the spalling of a layer of concrete that may extend over a distance of several reinforcing bars.

The appearance of a crack over a reinforcing bar provides the salt ready access to the steel. However, salt can migrate several inches through the concrete or asphalt without the presence of cracks, and penetration is largely dependent on the depth and permeability of the concrete. Roughly, for each additional inch of concrete depth, the chloride content of the concrete will be reduced by about one-half.

The introduction of studded snow tires in 1961 aided the pace of deterioration in roadway slabs. The resulting roadway wear is more severe and, hence, the protective cover over the steel is diminished.

INVESTIGATION OF THE UPPER LEVEL DECK SLAB

Structural Integrity of the Deck—1971

An investigation analyzed the deck for wear, cracking, concrete strength and the possibility of shear planes was completed in 1971. This investigation included a visual examination of the deck slab from the top, as well as coring of the deck and Windsor probe analysis.

This visual examination of the upper deck, which was completed in June, 1971, indicated the presence of 490 total defects in the deck slab, including patches, potholes, and exposed concrete areas.

It is important to note that while the visibly patched areas were numerous, they only accounted for about 1.5% of the total bridge deck surface. It was previously mentioned that 25.5% of the deck was repaired during the period 1961–70. Since the bridge was paved with asphalt in 1968, it is evident that 24% of the repaired deck was covered with asphalt and not visible to the naked eye.

The deck was tested for concrete strength and possible shear planes (delaminations) through the use of coring and Windsor probe analysis. The Windsor probe, which was developed by Port Authority engineers, determines the penetration resistance of hardened concrete by using a steel probe energized by a driving unit delivering a specified amount of energy. The penetration resistance of the concrete is measured by the exposed portion of the probe of a specified length. Under carefully controlled conditions, test results may be used to provide an estimate of compressive strength. It was the first time this test had been used on concrete where the surface was covered with asphalt. This adds an additional factor, which is possible variable thickness of the asphalt riding surface, to the variables known to exist in the procedure.

On the basis of this investigaton, the average compressive strength of the cores was determined to be 6,470 psi, while the probes showed an average compressive strength of 6,400 psi. The cores and probes indicated that not more than 0.5% of the cores would be under 4,200 psi. The data also indicated that less than one core in 10,000 could be anticipated to have a strength of less than 2,950 psi. For the probes, less than one in 10,000 would be expected to have a compressive strength of less than 2,700 psi. The average compressive strength of the concrete was therefore determined to be 6,375 psi, with approximately 0.5% of the total slab estimated to be under 4,000 psi.

A few of the cores taken contained part of a reinforcing rod and many of the cores had an imprint of a reinforcing rod or a bulb tee which showed evidence of rust. Some of this rust could be due to the original contact of wet cement with steel. In a few cases, the rust was more extensive than could be accounted for in this manner. In a particular case, the evidence of a horizontal crack and infusion of rust was very apparent. This evidence of rust, the imprint of reinforcing rods and bulb tees, and the generally short cores mostly at the level of the reinforcing rods suggested the existence of a horizontal fracture plane in the bridge deck. These suspicions were confirmed during the 1974 orthotropic deck test installation when delaminations and vertical cracks in the deck were observed in the exposed slabs for the first time. This horizontal delamination of the slab at the level of the top reinforcing steel was then estimated to be quite extensive on the basis of the appearance of the sawcut concrete slab sections that were removed from one 60-ft bay of the bridge deck.

Chloride Ion Analysis—1974

Chloride ion analyses were performed on concrete dust samples drilled from the deck in the Spring and Fall of 1974. The results of these tests are presented below:

AVERAGE FREE CHLORIDE ION CONTENT
MAIN SPAN—UPPER LEVEL DECK
(pounds/cubic yard)

	Depth (in.)	Lane 6 (2nd Lane from Curb)	Lane 8 (Curb Lane)
Fall, 1974	1	14.09 (high)	5.91
	2	13.85	4.58
	3	9.56	4.29
Spring, 1974	1	5.59	3.69
	2	3.48	3.04
	3	2.15 (low)	2.31

A chloride content of one to two pounds/cubic yard of concrete is generally considered the threshold level to initiate corrosion of the steel slab reinforcement.

Deck Slab Sounding—1975

The deck was sounded with hammers from both the roadway surface and the underside in September, 1975. Based on the inspection of 19,800 square feet from the roadway surface, the extrapolated deterioration of the Upper Level deck amounted to 117,500 square feet out of a gross area of 405,500 square feet. On the basis of the inspection of 16,450 square feet from the underside of the deck the deterioration was extrapolated to be 189,000 square feet. It should be noted that the two survey teams which did the inspection operated with complete independence and neither attempted to investigate the same panels nor to coordinate with the other. On the basis of this testing, it was estimated that somewhere between 29% and 46% of the Upper Level deck was in a deteriorated condition.

In addition to the sounding data, the inspection of the deck underside also revealed evidence of rust on the bottom flanges of the bulb tees along with the presence of efflorescence and stalactites on some areas of the concrete haunch. Also, there were numerous areas of transverse cracking, the crack spacing being about one foot on center.

RESTORATION/REPLACEMENT ALTERNATIVES

On the basis of the above data, it was concluded that the slab deterioration was extensive, accelerating, costly to repair, and warranted a comprehensive program of restoration or full deck replacement.

A study was therefore undertaken, which resulted in sixteen potentially viable schemes for restoration or replacement of the Upper Deck slab. These schemes were evaluated on the basis of the following design and operational criteria, which were established by the Port Authority Tunnels and Bridges and Engineering Departments:

—Required structural life of a minimum of 20 years, based on actual experience.
—Improved wearing surface over the existing pavement.
—All fourteen bridge lanes open to traffic during the peak traffic periods.
—A minimum of four traffic lanes, two in each direction, must be provided on the Upper Level during off-peak hours.
—Completion within two years.
—A median barrier is required between opposing traffic flows.

The sixteen schemes were divided into three groupings depending on the extent of demolition required. An evaluation matrix was used to compare the various schemes as to adherence with operational criteria, construction feasibility and cost of construction.

A brief description of the various schemes and evaluation follows:

Group I contained two schemes for epoxy injection of the existing slab and surface sealing with an epoxy or monomer membrane; five schemes for removal of the concrete to the top of the bulb tees and replacement with epoxy, or concrete-type materials; and three schemes for removal of the concrete to the tops of the bulb tees and welding of steel plates and/or grating to the bulb tees with or without concrete fill.

Group I schemes were rejected for the following reasons: 5½" of the bottom portion of the original concrete slab would remain with questionable integrity for the required 20 year life; some schemes would require long cure periods that could not be met; some of the materials proposed for use are relatively recent developments without sufficient experience to predict life; the welded connections of the plate and grating schemes would be of questionable value and difficult to check.

Group II contained three schemes which would require the removal of all existing concrete and rebars, except the bulb tees. These would be ground, sandblasted, and painted, and plates and/or grating attached by welding or bolting.

Besides the requirement for bridging over the construction area to facilitate erection and maintain peak hour traffic, these schemes would also require excessive welding and connections difficult to check as well as excessive and uneconomical manpower and equipment requirements to complete the work in two construction seasons.

Group III contained three schemes for complete removal of the existing deck slab, including the bulb tees, and replacement with an orthotropic steel deck or grating system supported on the existing secondary floorbeams.

In order that all lanes be available during peak hours, and the work completed in two years, two staging schemes were developed and explored. The first consisted of a bridging system that would be rolled into place to carry traffic over the construction area during peak periods; the second consisted of pre-sawcutting and jackhammer trenching of the concrete slab, while leaving the bulb tees intact and plating until removal of the slab at a later time in minimum 5 ft × 15 ft sections. Temporary modular deck panels would be available to cover any openings left in the deck upon removal of the slab sections and prior to installation of permanent deck panels. The second scheme, which favored the orthotropic deck, was considered the best staging scheme for replacing the deck while maintaining the eight-lane roadway during peak hours. The orthotropic scheme also would require less field time for installation and a greater reduction in dead load over the concrete-filled grating scheme. In addition, the orthotropic panels could be pre-paved in the shop, so that they could be safely used by traffic immediately upon installation.

It was, therefore, concluded that the deck slab should be replaced with an orthotropic steel deck, under a construction staging sequence requiring partial demolition and plating of the existing slab for peak hour traffic and with later removal of the slab in sections for installation of the prefabricated, prepaved orthotropic steel deck panels.

TEST PANEL INSTALLATION

In 1974, a test installation of the orthotropic scheme was conducted on a 60 × 44 ft section of roadway. Steel panels 11 × 30 ft and 11 × 60 ft of both open and closed rib design were installed by bolting to the existing CB16s, using the existing bulb tee rivet holes, after removal of the slab in 5 × 15 ft and 11 × 15 ft sections.

Three different pavements in combination with six different seal coats and four different bond coats were installed for testing. The pavements included asphaltic concrete, mastic asphalt, and epoxy asphalt. Joints between panels were either bolted or fully welded to evaluate both types of joint behavior. The test proved that the proposed method of deck replacement was feasible and economical, while the different structural designs, connections and pavement systems tested yielded valuable input to the final design.

UPPER DECK REPLACEMENT

Final Design

In early 1976, the final design for the orthotropic scheme was completed by Port Authority engineers and contract documents prepared. The plans included replacement of the eight-lane Upper Level deck of the bridge suspended span, an area of 405,500 sq ft, as well as replacement of the New York anchorage approach roadway, 28,500 sq ft, and the primary New York approach ramp, Ramp UX, an area of 38,000 sq ft. Both the suspended span and New York anchorage roadways were replaced with a steel orthotropic deck having an asphaltic concrete pavement, while the Ramp UX replacement consisted of a reinforced concrete slab with a latex-modified concrete wearing surface.

The project also included improvements to the Lower Level eastbound approach roadways in New Jersey. These modifications, completed by separate contracts prior to the start of construction for the Upper Level deck replacement allowed for increased utilization of the Lower Level eastbound bridge lanes during lane closures for work on the Upper Level.

Bids for the deck replacement contract were received on June 29, 1976 and the contract awarded to Karl Koch Erecting Co., Inc. of Carteret, New Jersey, with a low bid of $18,477,000. Construction began in March, 1977 and was completed on October 25, 1978, six weeks ahead of schedule.

Structural System

Four prefabricated deck panels, nominally 60 ft long by 11 ft wide, field-bolted together along three longitudinal joints, make up a typical deck module. This module is completely isolated from adjacent modules to allow for movements in the bridge and differential temperature expansion in the steel deck. Preformed neoprene compression seals are use to seal all joints.

The panels consist of a ⅝ in. thick deck plate stiffened by WT 7 × 13 ribs at 15 in. o.c. welded to the deck plate. Strap plates, 8 × ½ in. thick, having 1½ in Ø holes are welded to the underside of the tee flanges at the existing floorbeam locations. The strap plates are bolted to the existing floor beams, utilizing the rivet holes in the floor beam flange. Plate washers ⅝ in thick cover the two oversize holes in the strap plate to insure full bearing of the panel on the existing floor beams. Seven-inch angles are welded to the edge of each panel

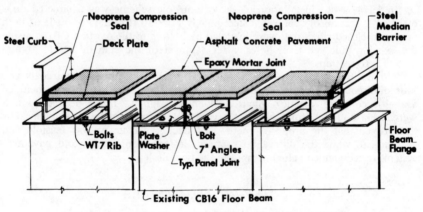

FIGURE 4. Typical section through orthotropic steel deck.

and bolted back to back connecting the individual panels together. There are no diaphragms between ribs except for two locations at approximately the mid-point of the module cross-section. These act as a zero point for transverse differential temperature expansion in the deck. The asphaltic concrete pavement is 1.5 inches thick (see FIGURE 4).

Material

All steel for the panels is ASTM A588 weathering steel, having a minimum yield of 50 ksi, and meeting the additional requirement for Zone 2 Charpy Notch Toughness. The deck panels are left unpainted, including the top surface of the deck plate, which is sandblasted to a commercial finish in preparation for application of the pavement tack coat. No protection of the surface was deemed necessary because of the dense nature of the pavement mix, as well as favorable corrosion behavior witnessed on battledeck and orthotropic deck roadway plates in the United States and Europe.

Pavement Selection

As a result of the field performance evaluation and extensive lab testing of various other materials, asphaltic concrete was selected as the pavement system. The major factors influencing the choice were:

—Satisfactory performance on the test panel.
—Ease of handling and application.
—Simplicity of future field repairs.
—Low initial first cost.
—Ten years of experience on other steel deck bridges.

The steel panels were prepaved off-site at the contractor's fabrication plant and when bolted into place on the bridge were ready to receive traffic immediately.

In the paving operation, four panels were bolted together in the final module configuration at the shop and paved as a unit. The pavement was then sawcut along the panel joint to provide a ¾ in wide joint and the panels separated for storage and eventual transportation to the bridge. After installation, the joint was filled with coal tar epoxy mortar. Testing of the 1½ in thick pavement was performed on the first set of panels fabricated under the contract. All possible conditions of handling were simulated and cores were taken to determine the bond to the deck plate. Based on this testing, only minor modifications were made to the pavement mix.

Construction Staging

Construction was planned with the eastbound traffic lanes redecked first, and the westbound the following year. In order to close half the Upper Level deck to traffic for the ten-hour night shift, a New Jersey-type concrete median barrier was installed on the north side of the bridge in the Spring of 1977, to separate the two-directional traffic flow. This barrier was then shifted to the south side of the bridge on the new orthotropic deck in 1978, for reconstruction of the north half of the bridge.

Since the steel orthotropic deck weight is only 60 psf or 43% less than the original deck slab, it was necessary that the deck replacement be staged longitudinally across the bridge in a pre-determined sequence to assure that no temporary construction condition would result in an overstress in any bridge member, including the tower stiffening trusses and wind bracing.

Construction Sequence

In order to ensure the erection of a minimum of four panels a night, which was essential to the two-year construction period, as well as ensuring that the deck would be ready to receive traffic each morning with an acceptable riding surface, the contract drawings showed a detailed construction sequence for slab demolition and panel installation, as well as details for roadway plating and temporary modular deck panels. Temporary modular panels are similar in detail to the permanent deck panels but smaller. A sufficient quantity of modular panels were present at the job site as a contingency to fill all possible combinations of openings in the deck created by the removal of slab sections.

The following sequence of operations was shown for panel installation (see FIGURE 5):

On the night or day prior to panel installation:

1. Install scaffolding below the deck and bust the rivet heads between the bulb tees and secondary floorbeams.
2. Sawcut the slab longitudinally for the full 60 ft bay length.
3. Jackhammer trenches longitudinally and transversely across the slab.
4. Plate over all sawcuts and jackhammer trenches.

On the night of panel installation:

1. Remove all roadway plates and burn the bulb tees.
2. Remove the slab by jacking free and lifting with the crane.
2a. Install temporary modular panels if for any reason the permanent panel installation cannot proceed (because of equipment breakdown, misfit of a permanent panel, or the onslaught of inclement weather).
3. Install the panel into position after completing the necessary preparatory work.
4. Install the bolts and open to traffic.

It should be noted that during the entire deck replacement, it was never necessary to install the temporary modular panels (step 2a above).

Erection

The contractor's scaffolding consisted of steel framing with welded steel decking. Scaffolding sections were hung from the existing bridge framing, with a total of five bays installed at any one time. These were moved by lowering onto a special truck frame, equipped with hydraulic lifts, which carried the section to a new location and lifted it into position for attaching hanger clamps. Scaffolding moves were made during off-peak hours with lane closures on the Lower Level roadway.

A 125 ton truck crane with a 90-ft boom was used by the contractor to remove slab sections and install new panels. This crane was able to completely replace the deck in a single bay without the need for repositioning. However, because of the heavy weight of the crane, special outrigger spreader panels and precise location of these on the bridge deck were required. With this crane, the contractor was able to remove the existing slab in 11 × 20 ft and 11 × 40 ft sections.

In order to ensure proper fit of the orthotropic steel panels, tight tolerances had been specified for overall dimensions and squareness of the four panel module as well as tolerances for flatness, straightness of the ribs and sweep. A paving jig that simulated the bridge framing was constructed by the contractor in the fabrication yard; sixteen panels (four modules) were assembled in this jig for the checking of tolerances and bolt hole locations, as well as for sandblasting, tack coating, and paving of the deck plate.

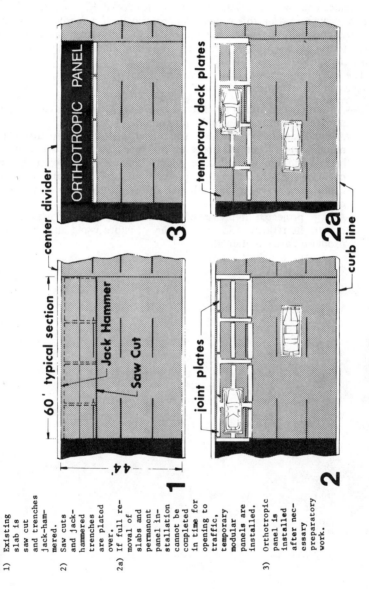

Major steps of
deck removal
and panel in-
stallation:
1) Existing
   slab is
   saw cut
   and trenches
   jack-ham-
   mered.

2) Saw cuts
   and jack-
   hammered
   trenches
   are plated
   over.

2a) If full re-
   moval of
   slabs and
   permanent
   panel in-
   stallation
   cannot be
   completed
   in time for
   opening to
   traffic,
   temporary
   modular
   panels are
   installed.

3) Orthotropic
   panel is
   installed
   after nec-
   essary
   preparatory
   work.

FIGURE 5. Construction sequence.

ADVANTAGES OF THE DECK REPLACEMENT SCHEME

This method of sawcutting and removal of a bridge deck in large segments and replacement with prefabricated units, developed by Port Authority engineers and pioneered on this project, has proved itself a feasible, cost effective, and speedy method of replacing bridge decks. This is especially true on heavily traveled structures where all lanes must be maintained during peak hours, as on the George Washington Bridge.

The orthotropic steel deck construction has the following additional advantages on this structure:

—An anticipated minimum fifty-year life with only repaving required at about ten-year intervals in the interim.

—A significant reduction in the total dead load of the structure (10.1%), reducing the level of stress in the main structural elements of the bridge and providing future reserve loading capacity. (This reserve capacity could be utilized for presently unforeseen loading conditions or could serve as a cushion should future degradation of the structure occur).

—And a projected improvement in the bridge's aerodynamic response due to the reduction in mass of the structure while all other structural properties are maintained.

PERFORMANCE, THE ULTIMATE JUDGE *

George Schoepfer

Of an audience comprising chiefly engineers, I am certain that the majority identify themselves more with construction than with maintenance and performance through the years. I would be the last to argue that painting and patching and cleaning and mucking are accompanied by anything like the sense of accomplishment associated with the timely completion of a major construction project. Don't we all revel in the memories of the times we were lucky enough to be intimately involved in the major projects that highlight our professional careers?

I plan to talk to you about bridges, more specifically long-span suspension bridges, and particularly about O. H. Ammann's bridges.

I do not present myself as a biographer of O. H. Ammann, for there are many in this audience and some on today's program who knew him much better than I. I did, however, get to know the "Gentle Genius" well over a ten-year period in the final and most productive years of his amazing career.

As you have been informed, I am the Executive Officer and Chief Engineer of the Triborough Bridge and Tunnel Authority. I started as an assistant in the bridge maintenance division some 23 years ago and assumed positions of increasing responsibility through the years as caretaker and owner's representative on new construction.

The history of suspension bridge construction, like the history of so many other activities, is marked by periods of intense productivity with several bridges under construction simultaneously and other long periods with no suspension bridge construction under way from coast to coast.

I was fortunate enough to be on the scene during a period of peak building activity, with two suspension bridges, the Throgs Neck and Verrazano-Narrows, under construction by the same Authority at the same time.

We are a society with some fascination for new construction, but little interest in day-to-day operations.

We grant awards and affix plaques for design excellence but not for excellent performance over the years or for decades of maintenance-free or minimum maintenance service.

Those of us who have ever been involved in maintenance have not been particularly aggressive or effective in promoting good structure performance or for that matter sending out warnings of unsatisfactory performance.

During periods when an agency is actively engaged in any substantial new construction, there is likely to be some owner-designer exchange of information simply because of the ongoing relationship between the parties. In the absence of a new construction program there is a tendency for designers to lose touch with the performance and behavior of their work.

We of the Triborough Bridge and Tunnel Authority have continued our

---

George Shoepfer is Executive Officer and Chief Engineer with the Triborough Bridge and Tunnel Authority New York, New York 10035.

* Luncheon speech.

0077–8923/80/0352–0157 $01.75/1 © 1980, NYAS

relationship with the designer through both our new construction and major maintenance program. We have all gained.

From experience at the Triborough Bridge we learned to avoid roadway joints over the anchorage structure at the Throgs Neck and Verrazano-Narrows Bridges to minimize water problems.

We learned to increase drainage openings and drain-pipe diameter and to avoid right-angle fittings.

We developed details for anchorage structures to provide sufficient ventilation to avoid perpetual dampness and at the same time fully enclose the structures to the extent that the pigeons are barred. To our own satisfaction, we dispelled the myth that a traveling maintenance platform is the best device for access and maintenance of bridge understructures. We have, of course, utilized a movable platform where necessary, but we have found a simple inspection walkway far more useful.

We have developed suspender connections totally visible for inspection.

A design detail that resulted in inaccessibility for subsequent maintenance is a cardinal sin. Mr. Ammann was well aware of this. Efforts to avoid this included full-scale plywood models of complex structural connections. As a further precaution, areas that we knew would be difficult to reach were given an extra coat of paint.

Some of you, I am sure, have gone through the difficult and expensive process of measuring and untorquing cable band bolts to verify bolt tension. We stenciled the unstressed length of the bolts directly on the bolt head and can confirm bolt tension at any time by a simple measurement.

But having said all of this, isn't it the actual performance of the structure that provides the final answer as to how well the designer has done his job?

This raises the complex question of how to measure performance. How do we rate Mr. Ammann's suspension bridges in this regard?

I suppose we must start by determining if they have fulfilled the transportation purpose for which they were built. Let us take a look at the Triborough, Bronx-Whitestone, Throgs Neck and Verrazano-Narrows Bridges. Total daily traffic 431,000 vehicles. Combined gross toll income in one year $141,886,-792.47. I would say that they are doing the job.

How do they blend with the metropolitan fabric of New York City? They are without doubt graceful, aesthetic additions to the City's day and nighttime skyline, considered by many to be mankind's most outstanding sculpture.

Are there elements of the structures that are outstanding from an engineering point of view? The fact that these bridges serve their intended purpose so well, as the traffic statistics demonstrate, and look so beautiful is the truest measure of successful engineering.

We point with pride to a number of things:

• Reinforced concrete roadways of the Triborough Bridge still in service with less than 30% resurfaced after 43 years of service.

• Battle deck steel plate pavement on portions of Triborough Bridge similar to orthotropic design in excellent condition after 43 years with one resurfacing.

• Structures that are maintainable and in fact have been maintained and stand ready to continue to function for the next 100 years or more.

For the world's longest suspension span, the Verrazano-Narrows Bridge, the design provided that the transverse floor frames, stringers, stiffening truss members and lateral bracing be so framed and interconnected as to form an integrated, continuous tubular framework particularly effective in resisting vertical, lateral and torsional forces on the suspended structure.

Aside from these generalities, is there anyway to measure true comparative performance of structures?

You know the bridge and tunnel business is a fascinating business. At Triborough we take in some $515,000 in small denomination bills and coins every day. It takes 35 people to sort and count it. Each week there is over $3 million to invest. But it also happens to be one of the most measurable businesses I know of. We have statistics on everything. We know the hourly count through every lane, and the number of vehicles each employee processed daily, weekly, monthly and yearly. I've earned the reputation of being something of a nut in trying to increase efficiency by measuring and increasing productivity. I am a director on the board of an international association of bridge, tunnel and turnpike operators, the IBTTA. Naturally, we all compare notes. I wonder if any of you have ever watched that TV show, "Name That Tune." The contestants say I can name that tune in five notes, another will claim to be able to do it in four notes, etc. Well some of our IBTTA sessions remind me of that program as I make the claim that I can process a car over a 2½ mile bridge for 13.8¢ and through a tunnel for 30.8¢ and challenge the group to top the figures. Or looking for challenge, I can boast that each toll collector at the Throgs Neck Bridge collects from 416,423 vehicles per year.

In playing this game, however, I have learned that it may be merely difficult or it may be impossible to obtain true comparative statistics. In the toll business there can be hidden forms of subsidy, an agency may or may not bear the expense of policing, work may be done by contract making maintenance payroll appear low. In fact, there are so many pitfalls in making statistical comparisons that I seriously doubt that we will be able to make any valid comparisons, but I do feel that the final measure or test of performance is to determine the cost to maintain the structure over the years. Obviously, on a major bridge structure the number of years studied is important. In a year during which a bridge is repainted or repaved, maintenance costs skyrocket. Taking all this into consideration, I think it would be worthwhile for us to look at our records and let the figures tell us how well our structures are serving us.

There is, of course, another problem in attempting such a comparison. Some of us over-maintain and others seriously under-maintain structures. I suspect that if we could determine structural maintenance cost on New York City's West Side Highway for its 40-odd-year life, the figure would be remarkably low, but it is also true that the structure collapsed several years ago.

In the case of my Authority, we do allocate costs and maintain accurate records of all expenditures. Although we do certain maintenance work with our own forces and other work such as repainting and repaving by outside contract, our records are all inclusive and all expenses are allocated.

Let we inject a thought about contract vs. in-house maintenance. In my experience this is a subject requiring periodic re-evaluation. We have realized dramatic savings by switching to contract services in certain areas, but contract costs can rise just as dramatically and it pays to check the relative advantages of in-house with contract services from time to time.

Getting back to overall maintenance expenses, I believe that an average maintenance cost over a ten-year period figured on a square foot basis gives us the best measure of performance of a structure. In the case of my Authority, the structures we maintain are toll bridges, and substantial maintenance effort and expense is related to toll equipment from the electronic recording equipment to the toll booths, plazas, and buildings. We have attempted to segregate these items to produce the following figures, which represent the ten-year average per square foot maintenance cost for the four O. H. Ammann suspension bridges under our jurisdiction:

	Maintenance Cost
• Triborough Bridge	$.68 per sq. ft. per year
• Bronx-Whitestone Bridge	$.92 per sq. ft. per year
• Throgs Neck Bridge	$.50 per sq. ft. per year
• Verrazano-Narrows Bridge	$.28 per sq. ft. per year

Are these figures worth anything to you? I would be most interested in exchanging information with anyone among you interested in comparing our costs with their own. While we credit ourselves with an efficient operation, we really have no comparative statistics to back up such an assumption.

I fully expect, however, that any comparison will confirm the fact that O. H. Ammann structures perform extremely well.

THE GOLDEN GATE BRIDGE: BACKBONE OF A TRANSIT SYSTEM

Frank L. Stahl

What is a successful bridge project? Is it one which overcomes long and stubborn opposition to its planning and major difficulties to its construction? Is it a bridge of outstanding beauty, in harmony with its surroundings? Or is it one that continuously serves the needs of the people by and for whom it was built? There can be no question that the Golden Gate Bridge qualifies under all these criteria as one of the most successful bridge projects ever conceived.

The story of the planning and construction of the bridge has been most thoroughly and eloquently told by its designer, Joseph B. Strauss, in what is known as the Chief Engineer's Report.[1] And nothing need be said here about the beauty of the bridge, long a national landmark and attraction to millions of visitors. What does need telling is the unique way this structure continues to successfully meet the demands of the population for which it was built.

EARLY HISTORY

What is now the City of San Francisco was first settled in 1776 by exploring Spaniards pushing north from Mexico and was named Yerba Buena. Merchant ships soon found the Bay an extremely attractive berthing and trading spot, and when gold was discovered in the Sacramento area, San Francisco quickly grew to the largest city on the West Coast. By the middle of the 19th century it became apparent that the City, hemmed in on three sides by water and by mountains to the south, would soon need better access to the north counties across the Golden Gate. That area was not only attractive for agriculture, industry and recreation, but it started to lure many new residents who daily commuted to work in San Francisco. Ferries provided an initial solution. As early as 1872, a bridge crossing the Golden Gate was planned, and then abandoned, by railroad builder Charles Crocker. With the invention of the automobile and the rapid increase of auto traffic during and after World War I, the Bay's ferry system became strained beyond its capacity.

The idea of bridging the Golden Gate was revived in 1916 by James Wilkins, a local newspaper editor, and San Francisco's city engineer M. M. O'Shaughnessy who started to inquire among engineers throughout the country about the feasibility and cost of such a project.[2] Only one engineer, Joseph B. Strauss, replied that a bridge was not only feasible but could be built at a reasonable cost. From then on, the name of Joseph B. Strauss was inseparably connected with the Golden Gate Bridge.

As Strauss would later remark, "it took two decades and two hundred million words to convince people the bridge was feasible." Strauss had soon prepared a preliminary design of a bridge radically different from accepted norms, combining the basic elements of a cantilever bridge and a suspension

Frank L. Stahl is Senior Associate with the firm of Ammann & Whitney, 2 World Trade Center, New York, New York 10048.

0077–8923/80/0352–0161 $01.75/1 © 1980, NYAS

bridge into a "symmetrical cantilever-suspension" hybrid, which, he estimated, could be built at a cost of seventeen and a quarter million dollars. Much more difficult, as it turned out, was to find general public backing of the project and means to finance and build the bridge. Strauss campaigned tirelessly for his bridge, picking up influential support votes in San Francisco and in the northern counties. By early 1923, an Association for Bridging the Gate was formed which petitioned the State Legislature for permission to form a legal district. The Golden Gate Bridge and Highway District Act of California became law on May 25, 1923. Among other things, the act provided for the financing of the bridge by the issuance of bonds secured primarily by tolls, but backed by the power of the District to levy taxes to make up for deficiencies in operating revenue. The taxing power issue provided a powerful argument for many land owners against the District. The fight against the bridge was joined by ferry interests who feared loss of their business and by lumbermen who were concerned with the influx of tourists. It took nearly six years before all legal obstacles were cleared; several counties declined to join the District. Finally, on December 4, 1928, the Golden Gate Bridge and Highway District was officially incorporated, made up of the counties of San Francisco to the south; and Marin, Sonoma, Napa, Mendocino, and Del Norte to the north of the Gate (FIGURE 1).

The District was, and still is, governed by a Board of Directors, its members appointed by the legislatures of the member counties. Joseph B. Strauss was named Chief Engineer of the District. However, his early design had drawn increasing opposition and the Board insisted that he engage as a board of consultants O. H. Ammann, whose record-breaking George Washington Bridge was then nearing completion in New York; Leon Moisseiff, an eminent theoretician who had worked with Ralph Modjeski on the Delaware River Bridge in Philadelphia and with Ammann on the George Washington Bridge; and Charles Derleth, a noted professor at the University of California's College of Civil Engineering. Soon thereafter, Strauss abandoned his earlier cantilever-suspension hybrid design and developed the true suspension bridge design as it now stands. It appears that Strauss' young assistant Clifford E. Paine was primarily responsible for this new development, but it must be assumed that Ammann and Moisseiff, on the basis of their earlier works, greatly influenced this decision.

Construction of the Bridge was officially started on January 5, 1933; it was completed and opened to traffic on May 28, 1937. Ironically, the wide-spread fear of, and opposition to the taxing power of the District—which had led to years of litigation and delays—proved to be completely unfounded. The last of the bonds sold to finance construction of the bridge was retired on July 1, 1971 without ever requiring resort to taxation of the member counties.

THE BRIDGE

In spite of efforts by competing ferries to hold on to their customers by lowering fares, the bridge proved to be an immediate success. In its first year of operation, 2.5 million vehicles were expected to use the bridge. Actually, 3.3 million vehicles crossed the span, a level not expected to be reached until 1941. From then on, and especially after the leveling off during the World War II years, traffic has increased steadily and dramatically. The original traffic forecast expected about 10.5 million vehicles in 1970.[1] The year 1970 saw nearly 33 million vehicles crossing the Golden Gate Bridge, and by 1978 traffic had passed the 36 million mark (FIGURE 2).

The Golden Gate Bridge is the only direct highway connection between San Francisco and Marin and the other counties to the north. It functions as a funnel for all vehicles moving between San Francisco and the north bay

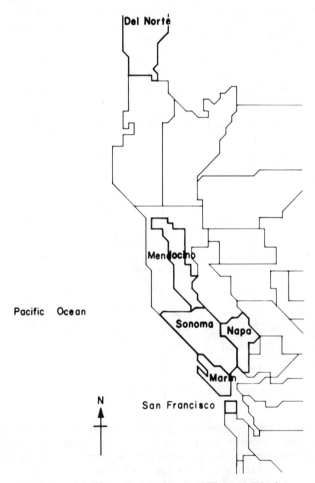

FIGURE 1. Golden Gate Bridge and Highway District.

counties. With a roadway of six 10-feet lanes, the tremendous number of rush-hour commuter vehicles can be accommodated only by a judicious switching of lanes several times daily to provide four lanes of traffic inbound to San Francisco in the morning and outbound in the evening. This arrangement is greatly facilitated by one-way city-bound toll collection.

The limited capacity of the bridge to accommodate continuing traffic growth has been the main concern of the District's Board of Directors and its staff for more than 20 years, and numerous studies have been made by the District, either in-house or in conjunction with other interested local agencies, in an effort to find ways to accommodate the ever-increasing traffic crossing the Gate.

The earliest study was made in connection with the development of plans for the Bay Area Rapid Transit System (BART) when an attempt was made to bring rapid transit trains across the Golden Gate Bridge into Marin County.[3] After investigation by the District's consulting engineers it was concluded that the bridge could not sustain the added loads imposed by the then contemplated type and weight of commuter trains, and the idea to extend BART into Marin county was dropped.

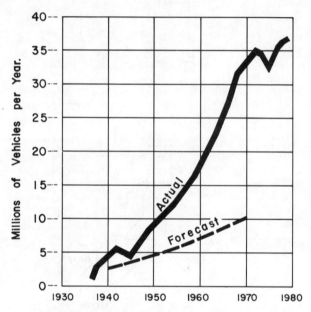

FIGURE 2. Traffic growth, Golden Gate Bridge.

By the early 1960s rush-hour traffic had grown to such proportions that it could not be accommodated anymore by the normal 3- and 3-lane arrangement. Fortunately, rush-hour traffic was—and still is—primarily one-directional, i.e. San Francisco-bound in the morning and Marin-bound in the evening. To ease congestion during the peak hours, a switch-lane operation was initiated in 1962, providing a fourth lane in the direction of heavy traffic flow by reducing the light flow direction to two lanes. Obviously, this could only be a short-range solution, and the search continued for other means to take care of the ever-increasing automobile traffic.

Over the next few years various studies were made by the Bridge District, the City of San Francisco, and the State—individually and jointly—resulting in many different proposals: a lower deck for vehicular traffic added to the present bridge [4]; a new parallel Golden Gate Bridge; a new bridge about one mile west of the present bridge, connecting the Presidio with Point Diablo in Marin County [5]; a new bridge about 3 miles east of the present bridge, from the Embarcadero area across Angel Island to Tiburon; a vehicular tunnel from Fisherman's Wharf to Sausalito; a rapid transit tunnel along the same alignment; public transit bus system across the Golden Gate Bridge; and high-speed ferry transit between downtown San Francisco and points in Marin County.[6]

With the retirement of the last bridge construction bonds scheduled for 1971, the continuing role of the Bridge District and its reponsibility towards the transportation needs of its member counties needed a redefinition. By an act of the State Legislature in 1969, the District was directed to develop a comprehensive transportation facilities plan and to provide or assist to provide transportation within the district. Thus, the Bridge and Highway District became a multi-modal transportation agency and its name was changed to the Golden Gate Bridge, Highway and Transportation District.

The District's Long Range Transportation Plan,[6] prepared by its staff and consultant, with input from a Citizens Advisory Panel and a Technical Advisory group, compiled and examined some 25 concepts of possible routes and transit modes for mass transportation in the Golden Gate Corridor (FIGURE 3). It confirmed that the travel capacity in the corridor was limited by the capacity of the Golden Gate Bridge, which, with four lanes operating in peak direction, is about 7,200 vehicles per hour. It also forecast that, with the increase expected for the next 50 years, U.S. 101 north of the bridge would require 18 traffic lanes by the year 2020 to carry the anticipated vehicular traffic if no transit facilities were provided. Even if such freeway and bridge construction was acceptable, the traffic load added to San Francisco's already congested streets would make this solution totally unacceptable.

In late 1970 and early 1971, 23 public meetings were held by the District throughout the corridor to consider the long-range transportation plans. It was obvious that public opinion was strongly opposed to the construction of new bridges or tunnels and highly in favor of the much less costly interim solution of improved bus and ferry transportation. Consequently, the District proceeded to establish an integrated bus and ferry transit system (FIGURE 4).

FERRY SYSTEM

The Bridge District's first step into the public transportation business was the purchase of a former San Diego sight-seeing vessel which was remodeled for service on San Francisco Bay. In August 1970, ferry service was initiated by the 15-knot, diesel-powered, 575 passenger vessel "Golden Gate" between Sausalito and downtown San Francisco (FIGURE 5). This marked the return of commuter ferries to the waters of the Bay which, in the ferry heyday prior to construction of the Golden Gate Bridge, had seen as many as 60 boats providing commuter service between its shores.

Simultaneously, engineering planning proceeded for the construction of three larger vessels as well as of permanent terminal facilities in Marin County and in San Francisco. Origin-destination surveys showed that approximately 50% of the commuters traveling to downtown San Francisco each day lived within a radius of 5 miles of the Central Marin County area of Corte Madera-Larkspur. In 1972, the Bridge District purchased 25 acres of waterfront property in the town of Larkspur for construction of its major Marin County

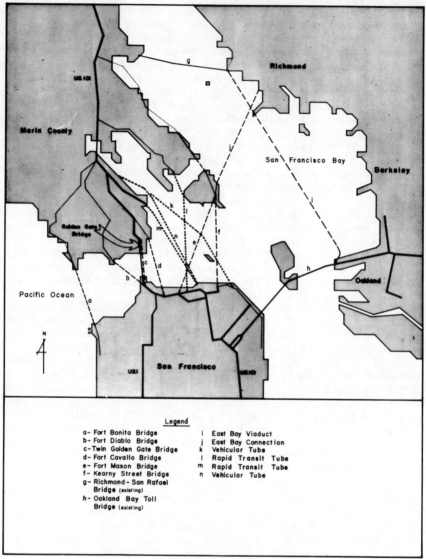

FIGURE 3. Proposed Golden Gate crossings.

ferry terminal. The Larkspur terminal, completed in 1976, has won wide architectural acclaim for its airy and functional appearance (FIGURE 6). An adjacent 950-car parking lot and feeder bus service is available for commuters. The terminal has docking space for three vessels, and includes integrated maintenance and overhaul facilities and a 300,000 gallon capacity fuel storage.

The downtown San Francisco terminal was completed in 1979 and is located within easy walking distance of the financial district and major business area.

Both terminals have been designed for complete compatibility with the new vessels. The multiple berthing facilities include water, fuel, sewage and electrical hookups within each docking system, which is designed to facilitate a speedy turn-around of the vessel. Shoreside gangways and ramps can be automatically raised and lowered to compensate for tidal variations and permit fast loading and unloading of passengers.

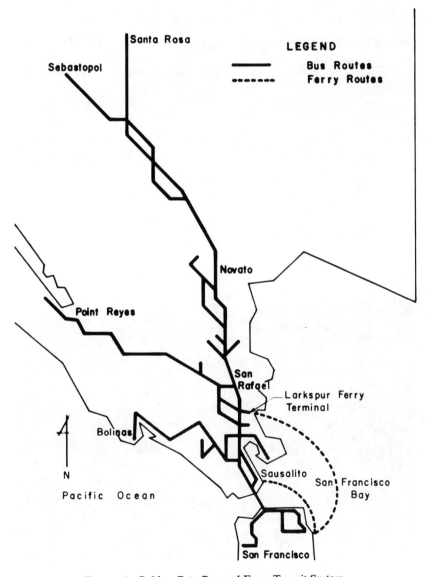

FIGURE 4. Golden Gate Bus and Ferry Transit System.

FIGURE 5. M.S. *Golden Gate*.

FIGURE 6. Larkspur Ferry Terminal.

Three new high-speed vessels each with 750-passenger capacity provide commuter service between Larkspur and San Francisco (FIGURE 7). In order to make the ferry system attractive to the commuter, the vessels have to be time-competitive with the automobile. The necessary speed of 25 knots is provided by three gas turbine engines coupled to waterjet propulsors. The engines occupy amazingly little space and can be operated independently of

FIGURE 7. G.T. *Marin* and Golden Gate Bridge.

one another thereby permitting high maneuverability and a large selectivity of vessel speed ranges. Engine rooms are unmanned; engines are controlled from the bridge and sophisticated electronic display panels alert the operator to any trouble area.

The hull design was chosen after extensive model testing. Lightweight marine aluminum alloy was used throughout its construction. In addition to speed, it was recognized that attractiveness and passenger comfort would be the key to public acceptability of the ferries. Great efforts were made to provide the passengers with amenities not usually found in public vehicles. Exterior and interior design and color schemes were coordinated for each vessel. Extensive use of carpeting and colored vinyl covering, large viewing windows, individual conversation areas with tables and comfortable first class seating, breakfast and cocktail lounges, sundeck and other amenities have indeed greatly contributed to making the District's new ferry system one of the most innovative, attractive and efficient public transportation systems in this country.

The success of the ferry system also has affected the area adjacent to the new Larkspur ferry terminal. A sprawling residential community and a large shopping center now occupy the site where 3 short years ago existed only a barren stone quarry.

BUS SYSTEM

Bus transit was inaugurated by the District in 1971 by taking over an ailing Greyhound commuter franchise and operating with leased equipment. With the purchase of its own buses (FIGURE 8), service was quickly expanded from purely local operations in Marin County to inter-county service between Marin and Sonoma Counties and trans-bay service to San Francisco. Today, the District owns and operates a fleet of 258 modern diesel-powered buses. This includes 10 articulated "bend in the middle" buses, which were acquired by the District as one of the first transit operators in this country for use on local routes in Marin County.

To lure commuters from their private automobiles, great efforts were made to assure the comfort of the bus riders. All coaches are air-conditioned. Airline-type reclining seats, covered in colorful fabrics, were installed and the

FIGURE 8. Golden Gate Transit bus.

FIGURE 9. San Rafael bus stop.

number of seats was reduced to 45 from the standard of 53 to provide more leg room. Bus routes were designed to reach into the neighborhoods of Marin and Sonoma where concentrations of commuters live, and to go into the San Francisco business and Civic Center district, where most of the commuters work (FIGURE 9). In addition, feeder bus routes connect many of these neighborhoods with the ferry terminals in Larkspur and Sausalito. Fares were set at a reasonable level and time schedules are adjusted on a continuing basis to satisfy the public's need. Full buses during all commute hours testify to the public's acceptance of the system.

The system includes attractive passenger shelters and directional information on many bus stops (FIGURE 10). The main terminal in San Rafael houses both the management force and a modern, fully integrated maintenance and repair facility for the entire bus system. Satellite terminals are located in Novato and Santa Rosa.

The effectiveness of Golden Gate Bus Transit was recognized and greatly improved by the California Department of Transportation, which allocated nearly 10 miles of freeway lanes in Marin County for exclusive bus use during commute hours. First, a counterflow bus lane for homebound commuters was taken from the opposite roadway for a 5-mile stretch just north of the Golden Gate Bridge through an area of frequent traffic congestion. Then, for an additional 4 miles, the fast lane was allocated for exclusive bus and carpool use during commute hours, southbound in the morning and northbound in the evening (FIGURE 11). This exclusive lane use helps to speed the trips of bus commuters, takes valuable minutes off bus schedules, reduces operating costs, and increases system efficiency by permitting more than one commute trip for many buses.

FIGURE 10. Bus directional sign.

FIGURE 11. Exclusive bus lane.

OTHER TRANSIT EXPEDIENTS

Other expedients to reduce the number of cars using the bridge during commute hours include free passage through the toll gate for cars with three or more occupants and free parking for bus riders in the District's parking lot in Mill Valley. Two other innovative transit programs are indicative of the pioneering spirit of the District.

Since 1971 the District has supported the formation of clubs for commuters who either live or work in areas not served by Golden Gate Transit. The District charters private buses on behalf of these clubs and subsidizes part of the cost of these charters; in addition, the District provides route planning and scheduling assistance. Currently, there are more than 20 club buses in operation

FIGURE 12. Van-pool vehicle.

providing more than 700 daily passenger trips and removing about 500 private cars per day from commute hour traffic. This club bus operation is a kind of pioneer service, getting people into public transit who would otherwise be unable to use it, and preparing the way for scheduled service in the future when economic conditions are more favorable for the establishment of mass transit connections.

Another innovative program is being carried out under a demonstration grant from the Urban Mass Transportation Administration (UMTA). Thirty-five vans of 10- and 12-passenger capacity were purchased by the District in 1977 and made available for commuters traveling to and from work between Marin, Napa, Sonoma, and San Francisco Counties (FIGURE 12). Under this plan, the van is leased to a driver-operator who assumes responsibility for organizing and maintaining a pool of passengers, with the assistance of the District's Van Pool Staff. The driver-operator plans routes, pick-up times and locations, collects fare payments, and takes over general responsibility for the operation of the van. The operator rides free and the passengers are charged a monthly rate which varies—depending on distance—from about 50% to 20% of the cost of operating a single-driver automobile over the same distance. Vans provide many advantages over normal passenger car operation: substantially lower cost per passenger mile; economical operation to and from areas of lower passenger concentration; personal pick-up service, and—since participants do not take turns driving as in the typical car pool—the possibility of dispensing with a second car.

EFFECT ON BRIDGE TRAFFIC

Despite the development of improved transit and the numbers of people riding in carpools and vanpools, traffic on the bridge still increases at an annual rate of 1.5 to 2%, down from 6% in 1969. When the Bridge District started its public transportation system in 1970/71, an average of 90,000 vehicles crossed the Bridge every day. Today, the average count exceeds 100,000 vehicles per day. Golden Gate Transit, however, by carrying more than 7 million transbay passengers annually on its buses and ferries, has removed the equivalent of about 18,000 cars from the daily commute traffic. Those 18,000 cars easily represent the difference between merely crowded rush-hour lanes and utter chaos. In addition, a considerable saving in energy and reduction in environmental pollutants has been realized through the public transit system.

Traffic studies made in 1969 set a goal of carrying 50% of all commuters by public transportation in 1980 and keeping the number of private cars at the 1969 level (FIGURE 13). Current figures indicate that vehicle growth during commute hours has been held to 10% over the last 10 years (FIGURE 14) and that 50% of the morning commuters are using public transit or carpools at least during the peak hour from 7 to 8 a.m. Efforts to meet the goal of carrying at least 50% of all commuters in public transit are continuing.

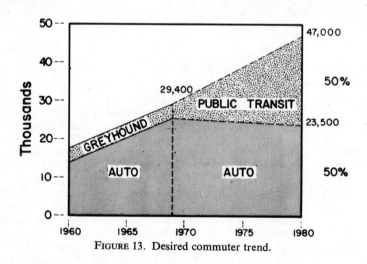

FIGURE 13. Desired commuter trend.

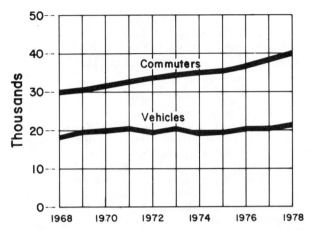

FIGURE 14. Commuter and vehicle growth—Golden Gate corridor (6-10 a.m.
South bound).

COSTS

At the beginning of the fiscal year 1971/72, when the last construction
bonds had been retired, the District had accumulated reserves of almost 23
million dollars earmarked for maintenance projects on the Bridge and for
creation of the public transit bus and ferry systems in the Golden Gate Corridor.

The cost of purchasing buses and ferries and of acquiring land and con-
structing the various bus and ferry terminals totalled about 61 million dollars.
Federal grants provided under the Urban Mass Transportation Act contributed
about 44 million dollars. The remainder came from, and depleted the District's
reserve funds.

The greatest challenge facing the District today is the need to provide the
necessary funds for the operation of its transit services. Unlike most other
transit organizations, Golden Gate Bus and Ferry Transit receives no local tax
support. Essentially, all operating and maintenance costs must be met from
current income of the District. Fare box receipts from bus and ferry service
pay for approximately 50% of these operating and maintenance costs, an
exceptionally high figure for a public transit system. The remaining money
must come primarily from the District's bridge tolls not needed for maintenance
and remedial work on the bridge itself. A small part is contributed by state
and federal operating assistance funds.

FUTURE GOALS

Even though there are now indications that previous long-range population
and traffic forecasts for the Bay area may have been on the high side, it is
realized that the present Golden Gate Transit System of buses and ferries has
capacity limitations. Some day in the not too distant future the current system
will not be sufficient to provide the necessary relief for the overcrowded bridge
roadway.

Therefore, studies must continue for transit improvements in the Golden Gate Corridor. Ideas being discussed include improved transit flow from the bridge into downtown San Francisco and exclusive busways with feeder lines in Marin County.

CONCLUSION

For the immediate future, however, the District's goal is the full realization of the capacity of its present transit system in order to keep traffic on the Golden Gate Bridge within manageable limits.

The bridge is and will be the centerpiece of the Golden Gate Transit Corridor. Neither the bridge nor the bus and ferry transit systems are, by themselves, able to provide the necessary commute capacity and are dependent on each other operationally and financially. Even though buses and ferries earn an unusually high proportion of their operating costs, the automobile commuter must assist through his bridge tolls to make up a large part of the transit system deficit. Thus, in this unique set-up, the automobile commuter not only pays for the privilege of driving his own car to and from work across the bridge, but by contributing to the transit operating fund, he helps others to keep their cars off the bridge by using public transit and thereby makes his own unrestricted use of the bridge possible.

REFERENCES

1. The Golden Gate Bridge. Report of the Chief Engineer to the Board of Directors of the Golden Gate Bridge and Highway District-California, September 1937.
2. CASSIDY, S. 1979. Spanning the Gate. Baron Wolman/Squarebooks.
3. PARSONS, BRINKERHOFF, HALL and McDONALD. 1955. Regional Rapid Transit (original BART report).
4. AMMANN & WHITNEY. 1968. Golden Gate Bridge Lower Deck for Vehicular Traffic.
5. STATE OF CALIFORNIA DIVISION OF BAY TOLL CROSSINGS. 1967. San Francisco—Marin Crossing.
6. KAISER ENGINEERS. 1970. Golden Gate Corridor—Long Range Transportation Alternatives.

THE SIGNIFICANCE OF DEFECTS IN WELDED LONG-SPAN BRIDGE STRUCTURES

Geoffrey R. Egan

Introduction

The trend to larger and larger bridge structures is brought about by the need to accommodate more people and to utilize size as a basis of increased efficiency. An increase in the load-carrying capacity of welded bridge structures, for example, usually means an increase in weight and section size or recently, a tendency to use of high yield strength steels. The latter may, in turn, lead to greater difficulties during fabrication, with a consequent higher risk of defects occurring during welding. Further, the use of increased section size makes it more difficult to detect and locate weld flaws using current inspection methods.

Major developments in welding technology since the mid-1940s have been primarily responsible for this trend to larger and larger welded bridge structures. As welding replaced riveting as a joining method, the ability of rivets to distribute load when cracking occurred and the ability of rivet holes to act as crack arresters no longer existed. In addition, there has been a general trend away from riveted ductile wrought iron and mild steel to welded higher strength alloy steels. These latter materials require the use of high strength weld deposits and are less tolerant of deviations from standard qualified welding specifications. It is the purpose of this paper to develop methods for assessing the significance of defects in welds in long-span bridges and to relate flaw size and material property requirements to the appropriate service conditions.

Potential Bridge Failure Mechanisms

Background

Before outlining potential failure mechanisms in bridges, it is necessary to develop a definition of what we mean by failure. In the extreme case, failure results in catastrophic collapse of the bridge, and the structure is completely unserviceable.[1] Before this event, however, various degrees of loss of integrity may occur. Even small amounts of crack growth can be considered as a "failure" because they may result in both a load restriction and an increased maintenance cost. In this sense, a failure can be defined as those events that lead to increased maintenance costs or reduced operating revenues (i.e., load restrictions). There are four predominant failure mechanisms which occur in welded long-span bridge structures. These are:

- Fatigue,
- Stress corrosion cracking,
- Brittle fracture,
- Plastic collapse.

Geoffrey R. Egan is with Aptech Engineering Services, Palo Alto, California 94303.

0077-8923/80/0352-0177 $01.75/1 © 1980, NYAS

Of these events, fatigue and stress corrosion cracking (SCC) are characterized by stable crack growth, whereas, brittle fracture and plastic instability are events which can lead to catastrophic collapse of the bridge structure. FIGURE 1 shows the general nature of these processes. Alternating stresses (fatigue, localized corrosion attack (SCC), or combinations of both phenomena (corrosion fatigue) cause crack initiation after some period of time, t_i. In most cases in welded structures, small intrusions at the toes of welds act as stress concentrators, and these serve as crack initiation sites so that the crack is present at time, $t_i = 0$. These cracks grow slowly during additional loading cycles or time, and final separation occurs when the cracks reach critical size. The critical flaw size may be determined from fracture mechanics principles for materials in which brittle fracture will occur and from plastic limit load methods when the material failure mode is ductile. Each of these failure mechanisms is discussed in detail below with particular reference to:

- The significance of weld defects and initial weld quality,
- Service stresses,
- Material selection.

Fatigue

Unlike fracture, which can occur without prior warning, fatigue is a slow process, and some time may pass after initiation of crack growth before cracking can be detected. Fatigue cracks are readily initiated from small (below detection level of current nondestructive examination (NDE) methods) pre-existing imperfections at the toes of welds, and a major portion of the total fatigue life is taken up in crack propagation.[2] Since crack propagation characteristics of all steels are similar, fatigue cracking of welded structures is not a problem that can be controlled by material selection. When the flaw grows to reach the critical size, unstable fracture (ductile or brittle) will occur, and the total fatigue life of a bridge will, therefore, depend on:

- The presence of any pre-existing flaws, as these reduce the time for crack initiation;
- The stresses, and the environment, in which the stresses are applied as these determine the amount and rate of stable crack growth;
- The fracture toughness or tensile properties of the material as well as the peak tensile stress acting on the bridge as these factors determine the critical flaw size at which final failure occurs.

To understand the parameters which may be used to design for adequate fatigue life, it is necessary to describe the fatigue process in applied mechanics terms.

The fracture mechanics parameter K, the stress intensity factor, which describes the local conditions of stress at a crack tip, can be used to quantify the rate of fatigue crack growth, da/dN, as follows:

$$da/dN = C_1(\Delta K)^n$$

were $\Delta K = K_{max} - K_{min}$ (i.e., the stress intensity factor range), and C_1 and n are parameters that depend on the material and the environment in which the structure is loaded.

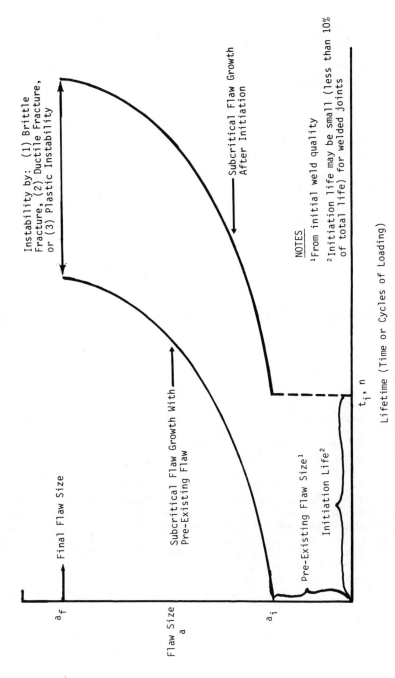

FIGURE 1. Schematic of flaw growth to failure with and without pre-existing flaw.

The cyclic stress, $\Delta\sigma$, is related to the stress intensity factor range, K, as follows:

$$\Delta K = C_2\, \Delta\sigma\sqrt{\pi a} \tag{2}$$

where C_2 is a constant to take account of flaw and part geometry. The rate of crack growth may, therefore, be expressed as:

$$\frac{da}{dN} = C_1(C_2\, \Delta\sigma\sqrt{\pi a})^n \tag{3}$$

This equation may be rearranged to obtain the integral from a_i, the initial flaw size, to a_f, the final flaw size as follows:

$$\int_{a_i}^{a_f} a^{-n/2}\, da = C_1 C_2\, \pi^{n/2}\, \Delta\sigma^n \int_{N_i}^{N_f} dN \tag{4}$$

From which the remaining life $(N = N_f - N_i)$ can be calculated:

$$N = \frac{1}{C_1 C_2{}^n\, \pi^{n/2}}\left[\frac{1}{1-\dfrac{n}{2}}\right]\left[a^{1-n/2}\right]_{a_i}^{a_t}\Delta\sigma^{-n} \tag{5}$$

From this equation, the remaining life, from an initial flaw size, discovered during a routine inspection can be predicted. In addition, this equation defines the parameters which have an influence on total fatigue life (i.e., a_i, a_f, $\Delta\sigma$, C_1, C_2, and n).

The Effect of Initial Flaw Size

The initial flaw size may be determined by: (1) the quality of the initial welds which are put into the structure; or (2) flaws that may have occurred by growth during service loading (an inspection, for example, would define a new set of initial flaw sizes for analysis). To illustrate the effect of initial flaw size, the remaining life has been calculated using (5) above for a range of initial flaw sizes, and a growth exponent of four, which is appropriate for most welded steel structures.[3] The results of this analysis are shown in FIGURE 2, where the remaining life is plotted as a function of initial flaw size. As an example, it can be seen that if the initial flaw size is increased from 0.05 inch to 0.1 inch, the remaining life reduces to something of the order of 50% of the life for a 0.05 inch initial flaw. That is, if the initial weld quality is such that the maximum flaw which escapes detection is 0.1 inch rather than 0.05 inch, half the life of the bridge has already been used up by permitting larger initial flaws to remain in the structure. For this example, the final flaw size is taken to be the same in all cases and equal to one inch. It can also be seen from FIGURE 2 that a threefold increase in initial flaw size from 0.1 inch to 0.3 inch reduces the remaining life to something like 25%. This indicates the strong influence of initial flaw size on structural life and illustrates the fact that it is essential to take account of initial weld quality when attempting to predict the fatigue life of welded bridges subjected to cyclic loading.

Effect of Final Flaw Size

Using the same equation and growth characteristics, it is possible to determine the influence of final flaw size on the fatigue life of a welded bridge structure. Using an initial flaw size of 0.15 inch, **(5)** has been used to determine the remaining life as a function of final flaw size, and these results are shown

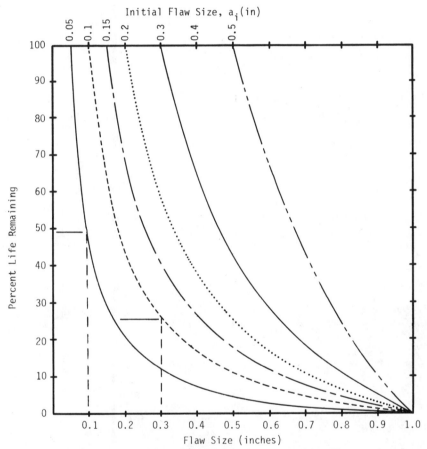

FIGURE 2. Effect of initial flaw size on remaining life.

in FIGURE 3. Note that a final flaw size ranging from one inch (that used for FIGURE 2) to five inches has been used. This figure shows that only marginal increases in remaining life are obtained for quite major changes in final critical flaw size. For example, an increase in final flaw size from one to two inches results in an approximately 8% increase in remaining life. The final flaw size has been determined from a brittle fracture criterion based on linear elastic fracture mechanics (LEFM). Once the applied stress intensity factor, K_I,

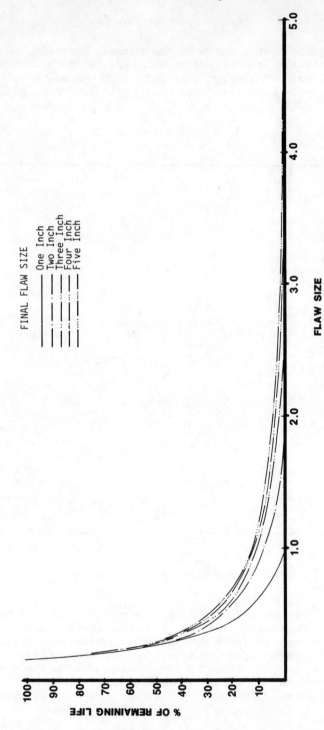

FIGURE 3. Influence of final flaw size on remaining life.

exceeds a critical value K_{IC} (the material toughness), fracture will occur. The applied stress intensity factor is determined as a function of stress level and flaw size from:

$$K_I = C_3 \; \sigma\sqrt{\pi a} \tag{6}$$

In this manner, it can be seen that major increases in fracture toughness or steel ductility have little effect on bridge lifetime when the controlling failure mechanism is fatigue. Further comments on this are made in the section dealing with fracture toughness and determination of final flaw size.

Effect of Cyclic Stress Level

Using (5), it is also possible to determine the influence of stress level on fatigue life. As an example, we assume that the cyclic stress has been reduced by a factor of two, and we want to determine the allowable initial flaw size which would give the same predicted service life. Equation 5 reduces to:

$$\frac{a_{i \; (1/2\Delta\sigma)}}{a_{i \; (\Delta\sigma)}} = \frac{16 \; a_f}{15 \; a_i + a_f} \tag{7}$$

and since $0 < a_i \le a_f$ numerical values of (7) range between one and 16. For high cycle fatigue in long-span bridges, $a_i \ll a_f$ so that a 16-fold increase in initial flaw size can result if stresses are reduced by a factor of two. Note that significant increases in allowable initial flaw size can result from minor changes in cyclic stress range since the initial flaw ratio is a function of the stress ratio to the fourth power.

It can be seen from the previous discussion that the important parameters which control fatigue life are:

- Initial weld quality through its influence on initial flaw size distribution
- Cyclic stress level.

It is also clear that the final flaw sizes determined by either fracture toughness or tensile properties have very little influence on total bridge life.[4] Since cyclic stress levels are determined by detailed design and loading conditions, it is important to control the initial weld quality to ensure that the predicted life times are achieved. Although we currently have a good understanding of the fatigue process in welded bridge structures, we still seem to be plagued with fatigue cracks occurring in bridge structures that have seen only a small part of their design lifetime. There could be several reasons for this:

(1) The cyclic stress levels are higher than anticipated.
(2) The number of loading cycles is higher than anticipated.
(3) The material has low fatigue initiation and propagation resistance.
(4) The weld quality is insufficient for the cyclic service loads.

Since not all bridges are affected, and (1), (2), and (3) are common to all bridges, it is unlikely that any of these are the primary cause of the early life fatigue cracking. It is the author's opinion that the primary cause is lack of initial weld quality; and without an improvement in weld quality, it may be necessary to use improvement methods, such as weld toe profile grinding, hammer peening, etc., to improve the fatigue performance of weld details. These methods are currently in use for offshore platforms and earth moving equipment.[5]

Stress Corrosion Cracking (SCC)

Although SCC in main welded bridge members is rare, SCC is more likely when (1) material yield strengths are high (i.e., of the order of 100 ksi), and (2) material carbon levels are high. In general, the lower the carbon content at a particular strength level, the less susceptible the material is to SCC.[6] Stress corrosion cracking may also be treated using principles of fracture mechanics that are used to evaluate fatigue crack growth. In addition, there exists, for most materials, a threshold of stress intensity factor (K_{ISCC}), below which stress corrosion cracks will not propagate. From the design point of view, it would be possible to determine the threshold level of stress intensity factor, in the appropriate environment, and then ensure that the initial flaw size and/or total stress level (dead load plus live load) results in K_I levels which are less than the critical value of K_{ISCC}. Since K_{ISCC}, the critical value of stress intensity factor below which SCC does not occur, is a function of σ and $a^{1/2}$, the influence of initial flaw size can also be determined. It may, however, be impractical to use this approach for materials that are susceptible to SCC. As an example, consider a material with a K_{ISCC} value of 10 ksi$\sqrt{in}$ [6] and a maximum applied stress of 30 ksi. The flaw size below which stress corrosion cracks will not grow in a susceptible material is given by:

$$K_{ISCC} = 1.2 \ \sigma_{max} \ \sqrt{a\pi} \qquad (8)$$

from which the critical flaw size to prevent SCC can be determined as:

$$a = \frac{1}{\pi} \left(\frac{K_{ISCC}}{1.2 \ \sigma_{max}} \right)^2 = 0.025 \ \text{inch} \qquad (9)$$

Since this flaw size is below a reliably detectable size, alternatives to inspection must be pursued to avoid failure by SCC. The most effective controls are material qualification procedures to identify SCC resistant materials.

Stress corrosion crack growth can be characterized using a fracture mechanics approach [7] in the form:

$$\frac{da}{dT} = C_4 (K_{max})^n \qquad (10)$$

For carbon manganese steels, values of n, in the range of six to eight, have been measured,[8] and this means that if the same life integration was performed as was done for the fatigue crack growth (Eq. 5), an even stronger influence of initial flaw size would result.

We do not have a large data base of SCC failures, primarily because bridge construction materials are not susceptible to SCC. It is likely, however, that in some environments the influence of corrosion will lead to accelerated cyclic crack growth rates and the phenomenon of corrosion fatigue. Corrosion fatigue can be treated in the same way as fatigue (previously described), except that the crack growth law should be based on accelerated growth rate data determined for the appropriate environment.

Fracture

A vast amount of research effort has been expended in recent years in attempting to characterize uniquely the resistance of a material to the onset of unstable crack propagation or brittle fracture. Many tests have been developed, and it now seems that through a study of the mechanics of fracture, a clearer understanding of the behavior of materials in the presence of a crack has been achieved. What is not so clear at present though, is how the principles of fracture mechanics can be used to provide an alternative, inexpensive and reliable method of assessing material toughness requirements for bridge structures, which do not warrant a full fracture mechanics analysis or fracture control procedure. Brittle fracture and elastic-plastic fracture have occurred infrequently in bridge structures; however, the focus of attention on the events that have occurred has tended to overemphasize notch ductility and notch toughness as a controlling parameter in bridge reliability. Brittle fracture, of course, is significant because:

- It has already lead to the catastrophic failure of several bridge structures.
- It can occur from the day that the bridge is under construction, even before service loads are applied.
- In most cases, there is virtually no prior warning of failure, and the fracture process is rapid.

Since the process of brittle fracture can be controlled by controlling initial flaw sizes, reducing stresses or increasing fracture toughness, it is possible to determine the significance of defects with respect to failure by brittle fracture. More recently, however, fracture control procedures have been evolved, which require only that the appropriate material is selected for referenced operating temperature ranges.[9] This fracture control philosophy assumes that errors and incorrect welding procedures, initial flaws, and other potential damage mechanisms during welding can be offset by the selection of appropriately tough material outside the area that has been altered by welding. These fracture control procedures will be described in detail in the following papers at this conference and will not be described here. Some particular points of interest will be highlighted as follows:

- As discussed above, initial flaw size limits for failure by brittle fracture will be much larger than those that necessarily must be controlled to avoid fatigue failure in the same structure. Therefore, in many cases, weld quality will be dictated by fatigue considerations.
- Current fracture control procedures take into account increased intermediate loading rates that occur in bridge structures, but do not take into account accelerated local crack tip loading rates that may occur, as a result of local damage and crack initiation from cracks in small brittle regions. An increase in fracture toughness, through material selection, is justified only if it allows better detection of fatigue cracks during in-service inspections. As discussed earlier, if a fatigue crack is discovered during service, it is likely that most of the remaining life of the detail has already been used up and additional toughness requirements will not lead to significant increases in fatigue life and, hence, bridge reliability.

Plastic Collapse

Very few failures in bridge structures are attributable to this form of failure. However, long fatigue cracks in ductile bridge steels will cause failure by ductile rupture and plastic collapse rather than brittle fracture. Well established limit load methods can be used to calculate the collapse conditions by accounting for the load carrying area removed by the crack. Interaction equations of the form:

$$\left(\frac{P}{P_\ell}\right)^2 + \frac{M}{M_\ell} = 1 \qquad (11)$$

will define plastic collapse limits.[10] P_ℓ and M_ℓ are the limit load and moment based on a flow stress (the average of the yield and ultimate strengths is used), and P and M are the applied load and moment. This method has been used for pressure vessels, pipelines, and beams.[10]

Weld Defects and Standards

The analyses performed using the principles of fracture mechanics have been carried out on the assumption that all defects behave as sharp cracks. This, in fact, is not the case, and most defects that are easily detected, such as porosity and slag inclusions, have a three-dimensional shape (rounded) and would be expected to perform much better than such defects as cracks, lack of side wall fusion, and lack of weld penetration, which are planar defects.[11]

Furthermore, not all weld defects will initiate growth immediately. This means that even if a weld defect is present, it takes time to initiate a fatigue crack, and the time to initiation is also a function of the loading spectrum (low stress cycles, following a high stress cycle, will result in a lower crack growth rate and retardation). Neglect of these effects, however, will lead to a conservative estimate of remaining life, and the methods outlined above can be used to make practical decisions on the significance of defects in welded long-span bridges.

If the major influence on bridge life is initial weld quality, what guidance do we get from current standards? TABLE 1 summarizes allowable flaw sizes from the AWS Structural Welding Code—Steel (AWS D1.1–79)[12] for:

- New buildings subjected to static loads only;
- New bridges, subjected to dead and live loads;
- New tubular structures, subjected to dead loads and live loads in a seawater environment.

Three forms of inspection—visual, radiographic, and magnetic particle inspection—are covered. For visual inspection, the allowable flaw limits are the same for all three structures irrespective of the loading type and environment. These are, of course, limits based on workmanship and do not have a "fitness for purpose" basis. For the other methods of inspection, the conclusion is essentionally the same. The allowable flaw limits are similar and relate only to flaw indication *length*. This is the only dimension that can be measured by radiographic and magnetic particle inspection methods. The dimension that controls the life of the detail, however, is the flaw *depth*. This is obviously an area where current standards could be improved by relating defect lengths to depths by a study of the aspect ratios of naturally occurring weld defects.

What can be done in bridge design to determine whether additional inspections, flaw limits, etc., should be specified? A simple screening procedure can be performed to determine "service critical" regions in the bridge structure. This assessment would be based on:

- Global stress levels,
- Detail geometry, complexity,
- Redundancy,
- Access and ease with which the detail can be fabricated,
- Weld procedures, repairs, etc.

Having identified these regions, a life integral can be performed using (5) to determine limiting initial flaw sizes for fatigue failure. This analysis should be based on linear cumulative damage and the relevant material crack growth data. Initial flaw limits can then be compared with flaw limits from standards to determine whether workmanship standards are adequate.

In-service Inspection and Flaw Evaluation

All of the previous discussion focuses on the significance of fabrication defects. The same methodology can also be used to evaluate:

- The need for repair,
- Adequacy of inspection intervals, and
- Bridge integrity

for flaws discovered during in-service inspection.

A strategy for performing these analyses is shown in FIGURE 4. It uses the fracture mechanics approach described earlier to predict stable crack growth. Stress and K analyses are combined with material properties to define the extent of crack growth and bridge integrity from NDE data.

Such bridge reliability improvement programs can lead to reduced long-term maintenance costs, particularly if the repair procedures include fatigue life improvement methods such as peening and grinding.

CONCLUDING REMARKS

Fracture mechanics analyses and the field experience indicate that there is a need for high initial weld quality so that early life fatigue cracking in welded long-span bridges can be avoided. Of particular importance is the use of fracture mechanics to define critical flaw sizes for failure by fatigue. The allowable or critical initial flaw size will depend on stress level, materials (as it affects the growth rate law constants), and to a minor extent, the fracture toughness or conditions that control final failure. This method enables the designer to determine whether adequate inspection procedures are available to detect the critical flaw sizes so calculated. If not, other strategies, which include fatigue life improvement methods, detail design changes, or a reduction in stress levels, must be developed to avoid early life fatigue cracking in long-span bridge structures.

TABLE 1

COMPARISON OF ALLOWABLE FLAW LIMITS (FROM AWS D1.1-79)

	New Buildings	New Bridges	Tubular Structures
Visual	No visible cracks allowed. *Piping Porosity* • Not allowed in *complete joint penetration groove welds* in butt joints if this defect is transverse to the direction of computed tensile stress. • For other *groove welds*, this defect shall not exceed $\frac{3}{8}''$ in any linear inch of weld and shall not exceed $\frac{3}{4}''$ in any 12" length of weld. • For *fillet welds*, the piping porosity shall not exceed $\frac{3}{8}''$ in any linear inch of weld and shall not exceed $\frac{3}{4}''$ in any 12" length of weld.	No visible cracks allowed. *Piping Porosity* • Not allowed in *complete joint penetration groove welds* in butt joints if this defect is transverse to the direction of computed tensile stress. • For other *groove welds*, the frequency of piping porosity shall not exceed one in 4" of length and the maximum diameter shall not exceed $\frac{3}{32}''$. • For *fillet welds*, same as for groove welds, except for *fillet welds* connecting stiffeners to web; here the sum of the diameters of piping porosity shall not exceed $\frac{3}{8}''$ in any linear inch of weld and shall not exceed $\frac{3}{4}''$ in any 12" length of weld.	No visible cracks allowed. *Piping Porosity* • Not allowed in *complete joint penetration groove welds* . . . (same as before). • For other *groove welds*, this defect shall not exceed $\frac{3}{8}''$ in any linear inch of weld and shall not exceed $\frac{3}{4}''$ in any 12" length of weld. • For *fillet welds*, the sum of diameters of piping porosity shall not exceed $\frac{3}{8}''$ in any linear inch of weld and shall not exceed $\frac{3}{4}''$ in any 12" length of weld.

| Radiographic and Magnetic Particle Inspection | In addition to the above requirements, a weld is unacceptable if:

• There are individual discontinuities having greater dimension of $3/32''$ or greater, and the greatest dimension of this discontinuity is larger than $2/3$ of effective throat, $2/3$ the weld size or $3/4''$.

Independent of the above paragraph, a weld is unacceptable if there are several discontinuities, each less than $3/32''$ in length, if the sum of their greatest dimension exceeds $3/8''$ in any linear inch of weld. | In addition to above, for welds subject to *tensile stress* under any condition of loading, the greatest dimension of any *porosity* or *fusion type* discontinuity that is $1/16''$ or larger shall not exceed the size in Fig. 9.25.2.1 (this has a range of $1/16'' \leq a \leq 1/2$.

For welds subject to *compressive stress* only, the greatest dimension of any *porosity* or *fusion type* discontinuity that is $1/8''$ or larger shall not exceed the size in Fig. 9.25.2.2 (range is $1/8'' \leq a \leq 3/4''$) | In addition to above, a weld shall be unacceptable if, given an individual discontinuity with a greatest dimension of $3/32''$ or greater, this dimension is larger than $2/3$ the effective throat or $2/3$ the weld size or $3/4''$.

Independent of the above paragraph, a weld shall be unacceptable if, given a group of discontinuities having a greatest dimension *less than* $3/32''$, the sum of the lengths exceeds $3/8''$ in any linear inch of weld. |

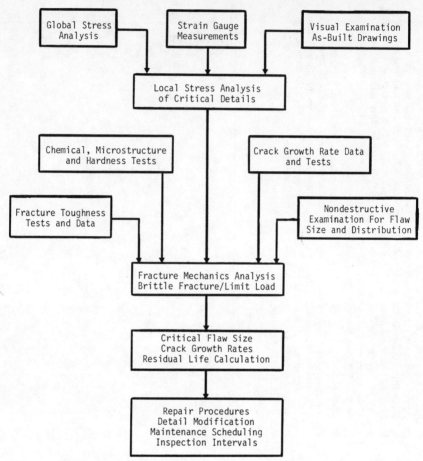

FIGURE 4. Flowchart of bridge repair and modification scheme.

SUMMARY

This paper describes methods of assessing the significance of defects in welded long-span bridges. The potential failure mechanisms that can occur in welded long-span bridges are also discussed. Analysis methods are developed to define the important parameters that control failure by fatigue, stress corrosion cracking, and fracture. Finally, the principles of fracture mechanics are used to determine the influence of initial weld quality on failure.

It is concluded that with current design procedures and allowable stress levels (dead weight and live load), workmanship standards (i.e., allowable weld flaw limits from existing standards) may be inadequate to prevent fatigue cracking during the design life of the bridge structure. To prevent increased failure rates due to fatigue cracking, it may be necessary to implement fatigue life improvement methods, such as weld toe grinding and peening as bridge restoration measures. Recommendations for bridge structural integrity evaluation are also made, which define the need for repairs and inspection intervals so that growing cracks can be repaired before fracture occurs.

REFERENCES

1. Report of the Royal Commission into the Failure of the West Gate Bridge, State of Victoria, Australia (1971).
2. HARRISON, J. D., F. WATKINSON & P. H. BODGER. 1970. The fatigue strength of welded joints in high strength steels and methods for its improvement. *In* Fatigue of Welded Structures. Tht Welding Institute (July).
3. MADDOX, S. J. 1969. Fatigue crack propagation in weld metal and heat affected zone material. Welding Institute Report E/29/69 (December).
4. EGAN, G. R., *et al.* 1976. An engineering analysis of the risk of brittle fracture of the Carquinez West Bridge. (July).
5. KNIGHT, J. W. 1975. Improving the fatigue strength of fillet welded joints by disc grinding the weld toe. Welding Institute Report E/60/75.
6. CARTER, C. S. & M. V. HYATT. 1972/73. Stress corrosion susceptibility of highway bridge construction steels. Federal Highway Administration Phase I and Phase II, Boeing Report No. D6159–1 (April 1972) and D6–60217 (April 1973).
7. EGAN, G. R. & R. C. CIPOLLA. 1978. Stress corrosion crack growth and fracture predictions for BWR piping. Joint ASME/CSME Pressure Vessel and Piping Conference, Montreal, Canada (June).
8. MCINTYRE, P. 1973. Influence of environments on crack growth in high strength steels. Conference on Mechanics and Mechanisms of Crack Growth, Cambridge, UK.
9. BARSOM, J. M. 1975. The development of AASHTO fracture toughness requirements for bridge steels. AISI Report (February).
10. MCNAUGHTON, W. P., G. R. EGAN & R. C. CIPOLLA. 1979. Significance of center-line lack of penetration in double welded stainless steel pipe—fracture analysis. Aptech Engineering Services Report AES-79–08–8 (August).
11. HARRISON, J. D. & J. DOHERTY. 1977. A re-analysis of fatigue data for butt welded specimens containing slag inclusions. Welding Institute Report 37/1977/E (April).
12. American Welding Society Structural Welding Code—Steel, AWS D1.1–79.

FATIGUE CRACKING IN LONGER SPAN BRIDGES

John W. Fisher and Dennis R. Mertz

Introduction

Long-span bridges, with the exception of the Silver Bridge failure at Point Pleasant, West Virginia have not exhibited serious problems with fatigue cracking and fracture. On occasion a stringer or floor beam has experienced cracking, which is a maintenance problem. When a main girder has cracked, the redundancy of the structure prevents collapse. Obviously, any cracked member is costly to repair and often restricts traffic until the repair has been completed.

In general, the fatigue-crack problems that have developed in longer span structures are not significantly different from those experienced with shorter spans. Cracking has occurred at poor details, at large weld flaws, and from secondary and displacement-induced stresses. In this paper we present three case studies to examine the actual behavior of several large structures where fatigue cracking developed.

One case examined is the cracking in stringer webs of a suspension bridge from out-of-plane displacement at floor beam connections. Strain measurements were obtained in order to assist with determining the cause of the cracking and to model the appropriate response. Further investigations were carried out on differently modified test bays in order to determine if the structural response could be altered so that further crack growth could be prevented or other courses or remedial action undertaken.

The final two case studies examine the cracking at details where large lack-of-fusion regions have resulted and permitted fatigue-crack growth to occur. In one case, lack-of-fusion conditions at groove-welded splices in longitudinal stiffeners were investigated. In the second case, similar undesirable conditions existed at lateral bracing connection plates that were welded to girder webs and transverse stiffeners. These conditions permitted fatigue-crack growth to occur in the transverse connection of the lateral plate and stiffener and into the main girder web.

The actual live-load stress range in longer span structures is generally of lesser magnitude and the frequency of occurrence is not as great as observed in shorter span bridges. Hence, it often takes longer for crack growth to develop in longer spans. Nevertheless, performance and behavior will eventually be impaired if fatigue-crack growth develops. Attention to details is shown to be of paramount importance in all bridge structures if fatigue and fracture are to be prevented.

Cracking at Lateral Bracing Connection Plates

On May 7, 1975, one of the main girders of the Lafayette Street Bridge over the Mississippi River in St. Paul, Minnesota, was discovered to be cracked.[1]

John W. Fisher is a Professor of Civil Engineering and Dennis R. Mertz is a Research Assistant at the Fritz Engineering Laboratory, Lehigh University, Bethlehem, Pennsylvania 18015.

0077-8923/80/0352-0193 $01.75/1 © 1980, NYAS

The crack was discovered in the east girder of the southbound structure in span 10. Span 10 is the 362-ft (110 m) main span of a three-span structure with side spans of 270 feet (82 m) (span 9) and 250 ft 6 in (76.4 m) (span 11). Forty-foot (12.2 m) cantilevers project into spans 8 and 12, as shown schematically in FIGURE 1. The web and flange were fabricated from American Society for Testing and Materials (ASTM) A441 steel. The crack developed in the girder web and the lateral bracing gusset plate. The crack occurred about 118 ft 8 in (36.2 m) from Pier 10. The crack had propagated in the web to within 7-1/2 in (190 mm) of the top flange when discovered. The entire bottom flange was fractured. The structure had been opened to traffic on November 13, 1968. The southbound lanes were closed from May 20, 1974 to October 25, 1974, for repairs to the deck and overlays. The estimated average daily truck traffic crossing the structure was 1,500 vehicles during the period November 1968 to May 1975. Thus, approximately 3,300,000 trucks had crossed the span at the time the fracture was discovered. A detailed study of the fractured portions of the web, flange, and gusset plate was made using material removed from the cracked girder. FIGURE 2 shows a schematic of the cross-section.

The material in the flange, web and gusset plates was evaluated to provide information on the chemical and physical properties of the plates. The physical properties of the flange and web plates are in good agreement with ASTM Specification A441. Both pieces exhibited good elongation and reduction in area characteristics. The gusset plate was in good agreement with ASTM Specification A36. The chemical composition of the flange and web plates were obtained by check analysis and is in reasonable agreement with the check analysis of the ASTM Specification (ASTM A441–63T, ASTM A36–62T).

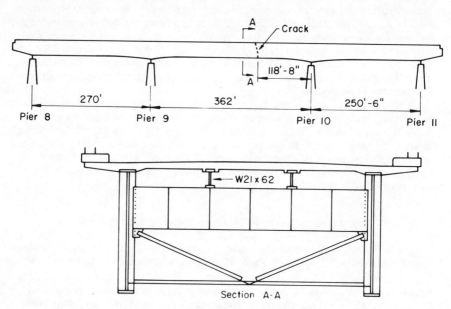

FIGURE 1. Schematic of span and cross-section of Lafayette Street Bridge.

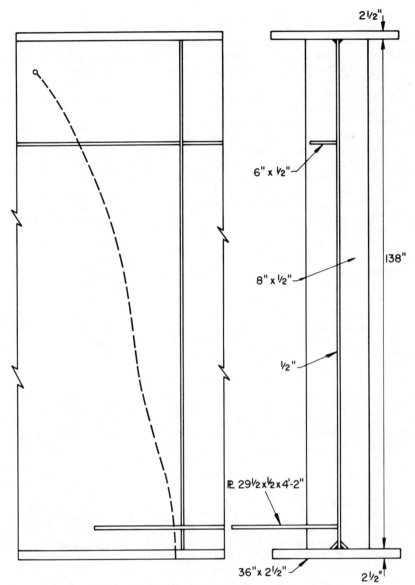

FIGURE 2. Schematic of Lafayette Street Bridge girder showing crack.

The flange and web plate show correspondence with the requirements of A441, and the gusset plate with A36.

The fracture toughness of the web, flange, and gusset plates were determined by making longitudinal Charpy V-Notch tests. A comparison of the web and flange A441 steel shows that the 15 ft-lb (20 J) transition temperature is about 50° F (10° C) for the flange and 15° F (−10° C) for the web. The gusset and flange exhibited nearly identical behavior. The flange plate meets the Charpy V-Notch requirements of the 1974 American Association of State Highway and Transportation Officials (AASHTO) Specification for temperature zone designation 1. [Minimum service temperature to −0° F (−18° C)]. The web plate meets the requirements for temperature zone 2 [−30° F (−34° C)]. The gusset plate material provided 15 ft-lb (20 J) at 55° F (13° C) and met the requirements for temperature zone 1.

The fracture surfaces indicate that the transverse and longitudinal single bevel groove welds that connect the gusset plate to the transverse stiffener and web did not penetrate to the backup bar at the root of the weld preparation. As a result, there was significant lack of fusion. Close examination indicated that the lack of fusion varied from about 0.15 in (3.8 mm) near the edge of the transverse stiffener up to 0.38 in (9.65 mm) near the web. Visual examination of the fracture surface indicated that fatigue crack growth had originated in the weld between the gusset plate and the transverse stiffener as a result of the large lack-of-fusion discontinuity in this location.

The fracture surfaces showed clearly that several stages of crack growth had occurred in the Lafayette Street Bridge. Both fatigue and brittle fracture modes were observed. FIGURE 3 shows schematically the apparent crack stages in the gusset-transverse stiffener weld and gusset plate, the web, and the flange. Stage 1 corresponded to the initial crack condition that resulted from the lack of fusion in the gusset-transverse stiffener connection. This resulted in a long semi-elliptical shaped crack which was about 0.38 in (9.65 mm) deep. Under normal truck traffic, fatigue-crack growth developed from this large initial crack condition and is shown as Stage II. Fatigue-crack growth occurred all along the gusset-transverse stiffener weld. Near the web, the crack was simultaneously drawn toward and into the web surface by the stress concentration provided by the longitudinal groove weld connecting the gusset to the web and the fact that the transverse weld intersected the stiffener weld and web creating a continuous crack path. The formation of the brittle fracture in the web corresponds to Stage III of the crack formation. No evidence of crack arrest was detected in the flange. This seems reasonable in view of the Charpy V-Notch test data. The fracture toughness of the web and flange were comparable, with the flange showing slightly less fracture toughness. The brittle fracture arrested in the web about 6 in (150 mm) above the gusset plate. At this point, there are no significant residual tensile stresses. Furthermore, the dead-load stresses were relatively small. In addition, the gusset plate was bridging the crack and prevented the crack from opening excessively. Following the formation of the initial brittle fracture, subsequent fatigue crack growth occurred in the web and in the gusset plate.

Final fracture of the web occurred as further fatigue-crack growth and yielding of the gusset plate occurred, which permitted the crack to open. This eventually led to the final girder fracture.

The fracture of a main girder of the Lafayette Street Bridge was due to the formation of a fatigue crack in the lateral bracing gusset to transverse stiffener weld. This fatigue crack originated from a large lack-of-fusion region that existed near the back-up bars. At the failure gusset, this groove weld intersected the transverse stiffener-web weld and thus provided a direct path for the fatigue crack to follow into the girder web. After the fatigue crack had nearly propagated through the girder web, it precipitated a brittle fracture of part of the web and all of the tension flange. Complete fracture was arrested by the lateral connection plate which bridged the crack. Subsequent cyclic load permitted further fatigue-crack growth over a limited time interval, which eventually resulted in substantial extension of the web crack.

A second smaller crack removed from the girder showed comparable fatigue-crack growth and verified the hypothesized crack development. The

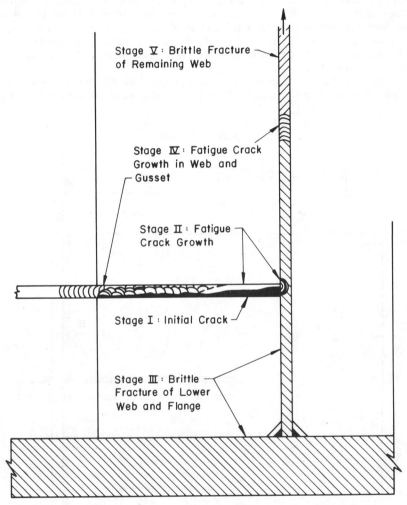

FIGURE 3. Schematic of stages of crack growth in gusset, web, and flange.

web plate was observed to just satisfy the fracture toughness requirements of the 1974 AASHTO Specification for temperature zone 2. The flange and gusset plates met the requirements for zone 1. If other fatigue cracks are removed or properly arrested, the material in the Lafayette Street Bridge will provide adequate resistance to brittle fracture.

The scheme shown in FIGURE 4 was used to retrofit the Lafayette Street Bridge. Vertical holes were cut into the gusset plate as shown schematically. After the gusset material was removed, the web surface and gusset could be carefully inspected for crack penetration using dye penetrant on the exposed web surface and the edge of the gusset. When any indication of a crack was provided by dye penetrant on the exposed web or hole surface, additional holes were cut into the girder web with a 1-1/4 in (32 mm) hole saw on the two diagonal lines from the opposite side of the web surface. These two holes intersect and removed an elongated section of the web. After these plugs were removed, the web surface in these holes were ground smooth and again checked with dye penetrant to ensure that the web crack did not extend below the area that was removed.

FIGURE 5 shows a suggested detail configuration for this type of connection, should it be used in the future. Both gusset plate welds should be prevented

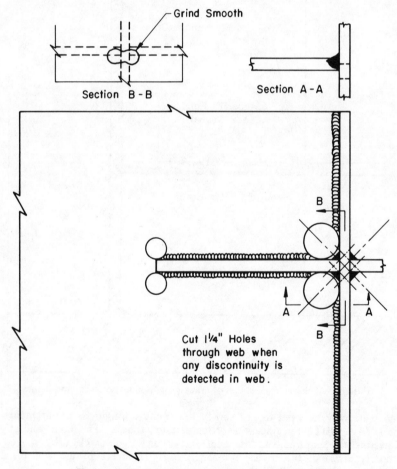

FIGURE 4. Preferred procedure for corrective action.

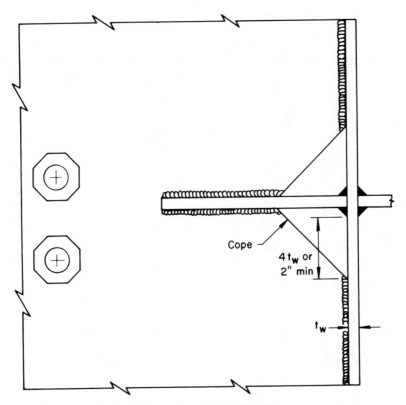

FIGURE 5. Recommended detail for future use.

from touching or coming too close to the transverse stiffener welds. This will prevent a path from forming into the web in case a crack develops in the transverse stiffener gusset weld. In addition, the gap reduces the restraint in the region. Furthermore, it would be preferable to not use a groove weld with backup-bar left in place that is perpendicular to the bending stress. These welds are likely to have lack of penetration and above average initial flaw sizes. As a result, they offer very low fatigue resistance and are likely to crack.

CRACKING AT GROOVE WELDS IN LONGITUDINAL STIFFENERS

In November 1973 a large crack was discovered in a fascia girder of the suspended span of the Quinnipiac River Bridge near New Haven, Connecticut.[2] FIGURE 6 shows the bridge profile. The crack was discovered approximately 34 ft (10.4 m) from the left or west end of the suspended span. The suspended span is 165 ft (50.3 m) long. The structure is noncomposite and the girders are 9 ft-2.75 in (2.8 m) deep at the crack location. FIGURE 7 shows the crack that developed in the girder web. The crack propagated to the mid-depth of the girder and had penetrated the bottom flange surface when discovered. The

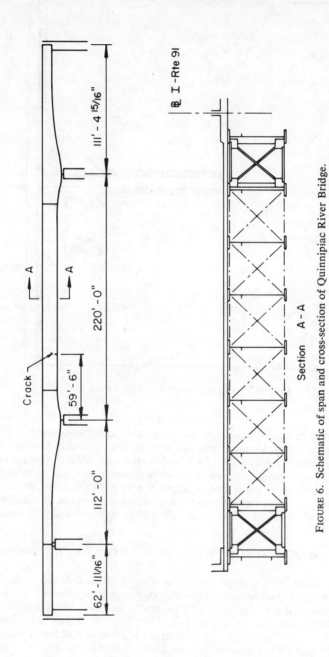

FIGURE 6. Schematic of span and cross-section of Quinnipiac River Bridge.

Section A - A

structure was opened to traffic in 1964. Thus, it had experienced approximately 9 years of service at the time the crack was discovered.

A study was undertaken to ascertain the causes of crack growth. To accomplish this task, half of the fracture surface was removed for visual examination. Examination of the fracture surfaces indicated that the fracture had started at the web longitudinal stiffener intersection. The fracture surfaces indicated that

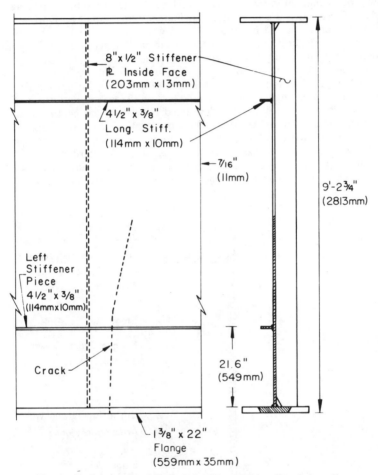

FIGURE 7. Schematic of Quinnipiac River Bridge showing crack.

a butt weld in the longitudinal stiffener had been made at this location but had never been completely fused. Close examination revealed that only a surface pass had been made and the reinforcement removed by grinding. Examination at the web-stiffener intersection confirmed that fatigue-crack growth had occurred. Fatigue-crack growth striations were detected adjacent to the longitudinal web stiffener break.

A detailed examination was also made of the fracture surface near the flange web intersection. No fatigue-crack growth striations were detected on the fracture surface of the flange. However, cleavage fracture was only observed to extend about 1 in (25 mm) into the flange. Also, the crack had obviously arrested and fatigue-crack growth was apparent from the macro-surface examination.

From the examination of the fracture surface it was apparent that crack growth had occurred in the Quinnipiac River Bridge in a number of stages and modes. These stages are illustrated schematically in FIGURE 8. During fabrication a very crude partial penetration weld was placed across the width of the longitudinal stiffener. It is probable that some crack extension from the unfused section occurred during transport, erection, and early service. Stage II of fatigue-crack growth would primarily develop after the stiffener was cracked in two. Stage III was the brittle fracture of the web during a time of low temperatures. It extended into the flange and up the web to mid-depth. Further fatigue-crack growth (Stage IV) developed thereafter and continued until the crack was discovered and repaired. This mode of behavior for Stages I and II was verified elsewhere on the structure. Other cracks were detected in comparable groove welds as well as crack extension into the girder web.

During the 9 years of service, the casualty girder sustained about 14.5 million cycles of random truck loading. It was estimated that between 2 and

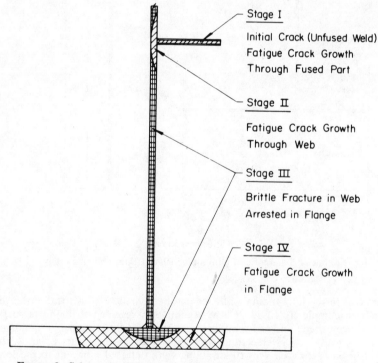

Stage I

Initial Crack (Unfused Weld)
Fatigue Crack Growth
Through Fused Part

Stage II

Fatigue Crack Growth
Through Web

Stage III

Brittle Fracture in Web
Arrested in Flange

Stage IV

Fatigue Crack Growth
in Flange

FIGURE 8. Schematic of stages of crack growth in stiffener, web, and flange.

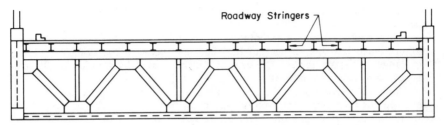

FIGURE 9. Schematic of suspension bridge cross-section.

20 million cycles would be required to propagate a ¼ in (6 mm) lack-of-fusion crack through the longitudinal stiffener thickness, depending on its proximity to a free surface. An additional 3.5 million cycles would be required to propagate the resulting crack through the web. These life estimates were in reasonable agreement with the traffic using the structure.

Crack instability appeared to develop when the net ligament of the web was finally broken. This resulted in a critical stress intensity of about 133 ksi$\sqrt{\text{in}}$. (143 MPa$\sqrt{\text{m}}$). The critical stress intensity factor was in reasonable agreement with the fracture toughness estimated from compact tension tests and estimates from Charpy V-Notch test data.

The random truck loading would have continued to increase the web crack through fatigue until a critical crack length was eventually reached even if higher fracture toughness was available.

Several other bridge structures are known to have cracked from comparable conditions.

CRACKING IN STRINGER WEBS OF A SUSPENSION BRIDGE

Cracking was detected in the stringer webs above the connection of lateral support brackets of a suspension bridge, as illustrated in FIGURE 9. As it was not altogether obvious why such cracks had developed, a study was carried out to determine the causes of the crack formation. Initial analytical models of the structural response did not provide a satisfactory explanation of the crack development. As a result, experimental studies were also carried out to assist in determining the causes of cracking and provide appropriate modifications to the analytical models.

In order to establish what steps were necessary to arrest crack development in other locations where comparable details existed, studies were also carried out on several proposed retrofit schemes. Two test bays were established, and modifications to the structure were examined to establish their adequacy. These measurements provided data on the traffic causing the high local stress condition, and permitted the effectiveness of several modifications to be determined by in-service conditions.

Out-of-Plane Bending Stresses and Cracks in the Stringer Webs

Visual observations of the fatigue cracks in the stringer webs had indicated that all cracks began at or near the fillets under the top flange of the steel beams at locations where lateral support brackets were attached to the stringers and floor beam trusses.

The presence of the lateral support bracket has permitted large stresses to be introduced into the stringer in the short gap between the bracket and the top flange of the stringer. Such stresses can be caused by rotation of the top flange and by relative out-of-plane movement of the web.

The field observations indicated that there were three major causes for the development of out-of-plane web displacement in the floor stringers. All three causes produced deformation in the short web gap above the bracket. Some rotation of the top flange was observed in the exterior beams during passage of trucks in these lanes. However, this displacement was not as significant as the lateral movement of the beam web. This was confirmed by both analytical and experimental studies. Following is a summary and discussion of the observed causes:

(1) Vertical displacement of the floor beam stringer at roadway relief joints occurs when gaps develop between the stringer flange and the floor beam truss from wear. Because the web offers little restraint to horizontal movement, very little shortening occurs in the bracket because the forces are small and large web movements occur. These deformations occur large numbers of times and have accelerated the crack formation and growth in the beams of the truck lanes.

(2) Shortening and rotation of the compression chord of the floor beam truss during passage of trucks also causes a relative movement between the slab and the top chord. Since the slab is very rigid in its plane, the shortening of the compression chord causes the bottom beam flange and bracket to translate as illustrated in FIGURE 10. The maximum effect of this shortening is greatest for the beams in the outside lane.

(3) The third and major cause of stringer web deformation is produced by the torsional response of the suspended floor system. This is illustrated schematically in FIGURE 11. The deformation of the torsion box composed of the floor beam trusses, the longitudinal trusses, and the lateral bracing system relative to the slab introduces large plate-bending strains into the stringer webs.

It was apparent from the first studies that only truck traffic produced large enough plate-bending strains to cause damage. Further studies confirmed that bus and light two-axle truck traffic could be neglected and indicated that the large strain variations were due primarily to five-axle trucks or grouped three- and four-axle trucks. If strains exceeding 75% of the maximum strain are assumed to cause damage, only about 13% of the three-, four-, and five-axle vehicles produced these "maximum strain" conditions.

Also of interest is the variation in plate-bending strain along the beam web. Typical of the change is the variation observed in an exterior beam, which is illustrated schematically in FIGURE 12. It is readily apparent that a rapid drop in plate-bending strain results away from the maximum restraint location. This was typical of the results observed at other locations.

An examination of the cracks that formed in the stringers indicated that the largest cracks occurred where gaps developed. In beams that were shimmed the crack would stabilize after it propagated into the low-stress-range region away from the gap.

It was apparent that substantially greater strains existed at the toe of the flange fillet, as that was where cracks initiated. Gages were placed at two locations on one side of several beams in order to measure the strain gradient in the web near the bracket.

FIGURE 13 shows the typical response of the gages on a stringer near the center of the slab at Panel 13 (in the side span). The strain gradient derived from the measurements illustrated in FIGURE 13 is plotted in FIGURE 15 (as gradient in as-built bracket). It is apparent that substantially greater strain exists at the fillet of the flange-web junction. The corresponding plate surface stress could easily exceed the yield stress because plastic flow was prevented by the adjacent elastic fibers and the steep strain gradient through the web thickness.

It should be noted that substantial variations in strain gradient between the

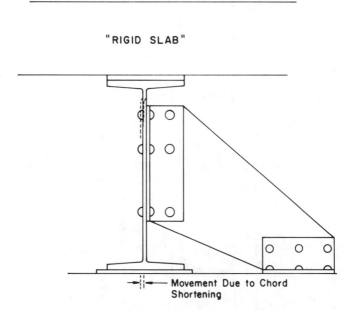

FIGURE 10. Schematic of influence of floor beam truss chord shortening.

web-flange fillet and the lateral support bracket are possible, depending on whether or not the bracket pushed or pulled against the stringer web. This had the effect of changing the restraint condition at the bracket.

The bridge tests at the AASHO Road Test and laboratory fatigue tests have demonstrated that the stress range is the only significant factor that governs the fatigue resistance of cyclically loaded systems.[3] Recent studies have also shown that Miner's Hypothesis can be used to estimate the damage developed in a randomly loaded bridge detail.[4] An effective stress range can be developed from Miner's Hypothesis and the exponential relationship between cyclic life and stress range. This results in

$$S_{\text{reMiner}} = \left[\sum \alpha_i \, S_{ri}^{3} \right]^{1/3}$$

where α_i is the frequency of occurrence of the stress range S_{ri}.

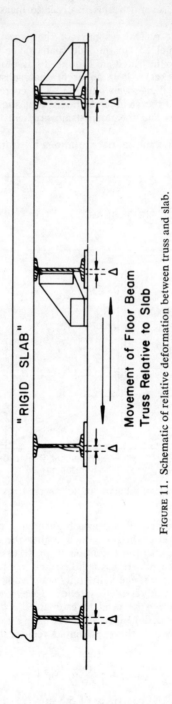

FIGURE 11. Schematic of relative deformation between truss and slab.

For the structure in question, about 3,600,000 three-, four-, and five-axle vehicles were known to have crossed the structure between the time it was opened in 1957 and when the cracks were observed in the late sixties.

The results of the measured out-of-plane plate-bending stresses extrapolated to the toe of the beam flange fillet are compared with the mean and confidence

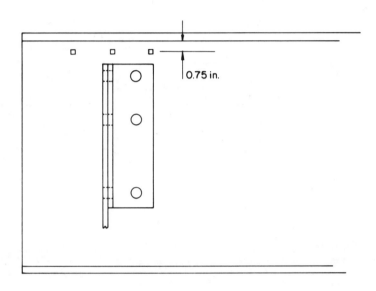

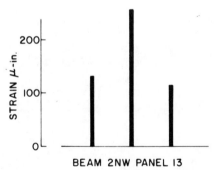

BEAM 2NW PANEL 13

FIGURE 12. Strain variation along beam web.

limits in FIGURE 14. The limits of dispersion were taken as twice the standard error of estimate given in ref. 5.

The test results demonstrate that the development of the web cracks was in accord with the expected fatigue resistance.

The test data acquired at floor beam trusses adjacent to the roadway relief joints also suggested those locations as well. Analytical studies suggested that higher stresses can be expected at these panels after cracks have occurred at the exterior panels of each bay.

Since torsion is one of the major causes of fatigue cracking, the lateral support brackets near the extremes of each bay are subjected to the largest torsional displacements. The brackets at the roadway relief joints have caused cracks to form, and the more flexible restraint that results permits larger deformations to occur in the adjacent panels.

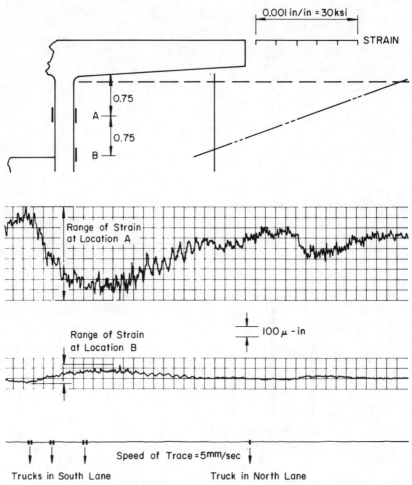

FIGURE 13. Strains on north face of Beam 8 SE, Panel 13.

On beams without lateral support brackets, relative deformation between the top and bottom flanges or rotation of the top flange does not introduce large curvatures into the beam web. Measurements on beams without brackets confirmed this.

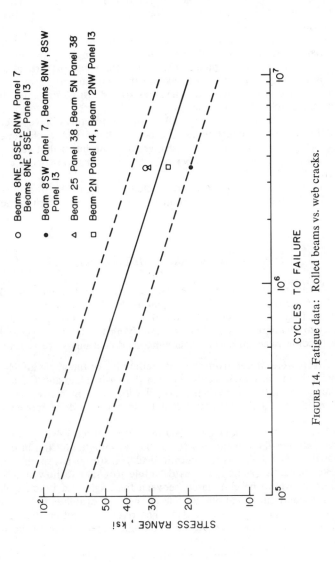

FIGURE 14. Fatigue data: Rolled beams vs. web cracks.

Fatigue Cracks in Bracket Angles

Fatigue cracks were also observed at many locations in one or both angles of the lateral support brackets. Usually, the web angle has a crack that begins at the top of the angle between the rivet and the angle fillet. When cracks developed in the angle attached to the floor beam truss they normally start at the edge of the rivet holes.

The lateral force bracket was designed to accommodate longitudinal expansion of the stringer. The ⅜-in plate attached to the angles provides very little restraint to longitudinal movement. This is not the case for vertical and horizontal displacements. The rivets connecting the components form an effective rigid joint. Hence, when clearance develops, the deformation must be transmitted into the bracket and beam web.

Most of the fatigue cracks and all complete fractures have occurred in angles in the line of stringers under the outside lanes where clearances have developed between the bottom of the stringer and the fillers that rest on top of the floor beam truss. A large number of cyclic deformations occur once these gaps develop.

At locations where the stringers are riveted directly to the floor beam truss, no vertical deformations can be introduced and very few fatigue cracks have formed.

Results of Bracket Modifications

Several minor modifications were made to the lateral support bracket near the middle of the roadway, in the side span at Panel 13, and their effectiveness was evaluated by strain measurements at that point. The purpose of these studies was to evaluate various ways of modifying the lateral support brackets should it become necessary. The following modifications were examined.

(1) Rivets were driven from the outstanding leg of the angle attached to the beam web and any tack welds between the ⅜-in plate and the angle were broken. The rivets were then replaced with ¾-in high-strength bolts and lock nuts. The bolts were installed loose and could be moved by hand in the oversize ¹⁵⁄₁₆-in drilled holes.

(2) The loose-fitting bolts described in (1) permitted relative movement to occur between the outstanding leg of the web angle and the bracket plate. Since tightened bolts would permit load transfer between the angle and bracket plate by friction, it was desirable to evaluate how much relative movement would be eliminated between the angle and bracket plate.

(3) A second modification considered was to evaluate the influence of bolting the angle to the beam web. The rivets connecting the angle to the web of a beam were removed and the angle was fastened to the web with bolts.

(4) Since one of the major reasons for the large plate-bending strain in the beam web was the short length of beam web between the bottom of the top flange and the top of the web angle, it was logical to attempt to increase the available length of beam web. To accomplish this modification, the top of the web angle was cut off so that only the two bottom bolts could be used to attach the angle to the stringer web.

The two basic bracket modifications, i.e., providing loose bolts in the outstanding angle leg (which is analogous to removing the bracket), and cutting the web angle short can be evaluated by considering Figures 15 and 16.

FIGURE 15 compares the strains and strain gradients of the two bracket modifications with the results obtained prior to the change. It is apparent that the strain gradient was not affected by the loose bolts. Only the magnitude of strain was decreased.

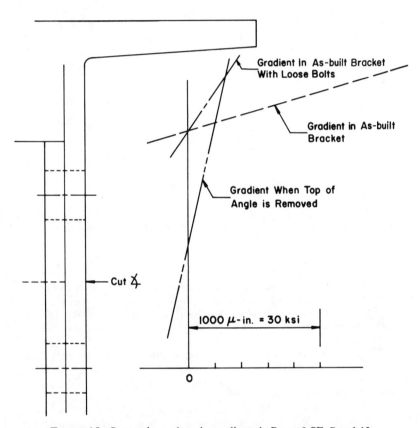

FIGURE 15. Comparison of strain gradients in Beam 8 SE, Panel 13.

When the top of the angle was cut off there was a decrease in the measured plate-bending strains, as indicated in both FIGURES 15 and 16. However, more significant was the reduction in the strain gradient. The actual strains at the toe of the fillet was about the same for both bracket modifications.

Bolting the angle to the beam web did not produce any difference in behavior, as was expected, since the web was still pushed out-of-plane by the torsional deformation. Tightened bolts caused a variation in strain range equal to about ⅓ the variation observed prior to the change.

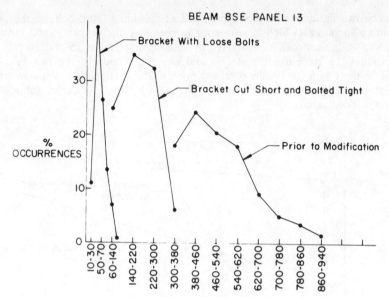

FIGURE 16. Typical strain history for bracket modifications.

Evaluation of Other Modifications to the Stringer Floor Beam Truss Connections

On the basis of the minor modifications that were made to the centerline bracket, two conditions were examined in greater detail at a typical bay. A "rigid" shear diaphragm was added at each end of the bay at the expansion joints. This modification provided an opportunity to determine whether the displacement could be reduced enough to prevent the undesirable movement from developing.

A second study was carried out after removing all lateral support brackets from the end two floor beam trusses adjacent to the roadway relief joints. It was desirable to determine whether the increased flexibility would be harmful to the remaining brackets between joints.

One test bay was established in the suspended span between Panels 43 and 49 to explore the effectiveness of installing a "rigid" diaphragm. Analytical studies had indicated that installation of a bracket would provide some relief. The analysis considered the shear diaphragm to be an elastic member—not a perfectly rigid one. The "rigid" shear diaphragm must be capable of resisting a substantial portion of the relative deformation that occurs between steel filled grid floor and the floor beam trusses.

Analytical studies indicated that a bracket's effect was confined to the end of the bay where it was installed. The analysis indicated that a bracket at the opposite end of a bay had no effect on the displacements and web moments at the other end. In view of this analytical finding, it was decided to install the "rigid" shear diaphragms at one end of the bay only. This would permit the acquisition of directly comparable information at a roadway relief joint.

The diaphragms were installed on each side of the roadway relief joint at Panel 49. FIGURE 17 shows the diaphragm that was used. The lateral support brackets were left in place when the "rigid" diaphragm was installed. This permitted a direct evaluation of the effectiveness of adding additional restraint to the torsional displacement that occurs at a roadway relief joint.

A second test bay was established between Panels 43' and 49'. Its purpose was to evaluate the effectiveness of releasing the lateral support brackets at the roadway relief joints and at other interior panels as necessary.

The bracket release was accomplished by removing the rivets in the outstanding legs of the web angle that connected the angle to the lateral brace

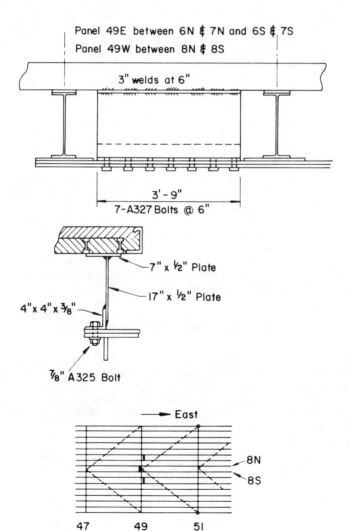

FIGURE 17. Schematic of "rigid" diaphragm.

plate. All tack welds were broken. This permitted relative movement between the angle and the plate.

Initially brackets were released at the roadway relief joints at Panels 43′ and 49′. After the response was obtained brackets were also released at adjacent Panels 44′ and 48′. No further releases were introduced, as the measurements and analytical studies indicated that they were not required.

Results of "Rigid" Shear Diaphragm

The effectiveness of the "rigid" shear diaphragms that were installed at Panel 49 was examined by strain measurements. Two bays were examined at Panel 49; the west bay between Panels 43 and 49 and the east bay between Panels 49 and 55. This permitted an evaluation at each end of a bay, so that a fixed-end and an expansion-end could be examined.

Measurements were taken in: (1) the as-built state; (2) with diaphragm installed on the west side of Panel 49, but in the as-built condition on the east side of Panel 49; and (3) with diaphragm installed on the east side of Panel 49, but with no restraint except the as-built condition on the west side of Panel 49.

The installation of diaphragms at the roadway relief joint was not as effective as analytically predicted in decreasing the plate bending strains in the stringer webs.

The diaphragm reduced the plate bending strains in the interior stringers at the roadway relief joint by 5 to 25% (see FIGURE 18). No reduction was observed at the next adjacent interior panel.

The local response of exterior stringers was not affected by the "rigid" diaphragm, although about the same reduction due to the torsional response was observed.

Installation of a "rigid" diaphragm did not appear to be a satisfactory means of reducing the plate-bending strain at roadway relief joints nor preventing the formation of cracks at the next adjacent interior panel.

Results of Releasing Lateral Support Brackets

Three response conditions were examined at Test Bay 2, which was composed of the section of the suspended span between Panels 43′ and 49′. This included: (1) examining the as-built state which would form the boundary conditions for comparison with any subsequent modifications; (2) releasing the brackets at Panels 43′ and 49′; and (3) releasing the brackets at the adjacent Panels 44′ and 48′.

Releasing brackets at the end panels was effective in reducing the plate-bending strain at the roadway relief joints. This was accompanied by an increase in plate-bending strain in the beam webs at adjacent floor beam trusses. This increase could lead to eventual cracking of the beam web as the measured strains suggested stress ranges as high as 33.0 ksi (228 MPa) developed. Although this crack formation is probably not critical, the effort to remove the end-panel brackets only would not justify this eventual crack formation.

The removal of lateral support brackets at the roadway relief joint and at the next adjacent interior floor beam truss is a satisfactory means of arresting further crack development at the roadway relief joints and preventing cracks elsewhere in the structure. The removal of these brackets reduced the plate-bending strain in the stringer webs to satisfactory levels, so that existing crack growth is terminated and the possibility of crack initiation at other locations was eliminated.

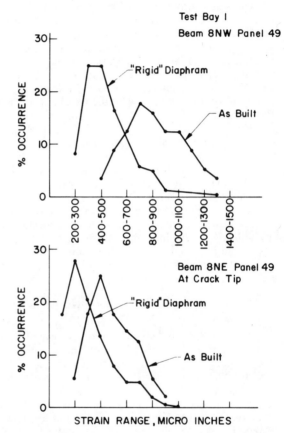

FIGURE 18. Histogram of strain response in Test Bay 1.

CONCLUSIONS

Long-span bridges, when compared with those with relatively shorter spans, have not generated serious fatigue cracking and fracture problems. However, despite lesser magnitudes of live-load stress range and lower frequencies of occurrence, (several vehicles may only produce one stress cycle), performance of long-span bridges has on occasion been impaired by fatigue cracking. The growth of these cracks has occurred over substantially longer periods of time as a result of the lower live-load stress.

Cracking has been shown to have occurred at relatively poor details, at large weld flaws, and at regions of structural components sensitive to out-of-plane displacement-induced stress.

Through the example of the Lafayette Street Bridge, fatigue cracking from an unfused weld at a relatively poor detail has been illustrated. To prevent such failure of primary structural members, the formation of a path into the major structural components in the event of crack development must be circumvented. Further, reduction of restraint must be achieved in regions of potential weld flaws. Through more careful design of details with regard to fatigue cracking these objectives can be reached.

Cracking in the Quinnipiac River Bridge illustrates further problems created by relatively large initial weld flaws. A lack-of-fusion condition in a very crude partial penetration weld across a longitudinal stiffener caused fatigue cracking to begin at very low levels of stress. Problems such as this can be prevented by more careful quality control of the welding processes and adequate nondestructive inspection.

Finally, one of the problems of fatigue cracking due to displacement-induced stresses was presented through an investigation of cracks in the stringer webs of a suspension bridge. The solution of this type of cracking problem consists of the elimination of displacement-induced stress by the use of joint details that do not impose large stresses in short gaps.

Through these three case studies we have seen that, while crack growth in longer span bridges may require substantially more time to develop because of reduced stress ranges and frequencies of occurrence, such problems do exist. By utilizing the techniques discussed the fatigue and fracture problems can be controlled if not eliminated. Attention to details is shown to be of paramount importance in all bridge structures if these problems are to be prevented.

ACKNOWLEDGMENTS

This paper is based on studies of several cases of fatigue cracking that developed in in-service structures. The authors are indebted to the Minnesota Department of Transportation, the Connecticut Department of Transportation, and the Delaware River Port Authority for providing details and information.

Thanks are also due Professors A. W. Pense, R. Roberts, and G. R. Irwin for their assistance and suggestions. Mr. H. T. Sutherland, Fritz Engineering Laboratory Instruments Associate, made possible the acquisition of field test data. Thanks are also due Mr. R. Sopko for a photograph, Mr. J. Gera for preparation of the figures, and Mrs. Ruth Grimes for typing the manuscript.

REFERENCES

1. FISHER, J. W., A. W. PENSE & R. ROBERTS. 1977. Evaluation of Fracture of Lafayette Street Bridge. Journal of the Structural Division, ASCE **103** (No. ST7): 1339–1357 Proc. Paper 13059, July 1977.
2. FISHER, J. W., A. W. PENSE, H. HAUSAMMANN & G. R. IRWIN. Analysis of cracking in Quinnipiac River Bridge. Submitted to the Journal of the Structural Division, ASCE.
3. FISHER, J. W. & I. M. VIEST. 1964. Fatigue life of bridge beams subjected to

controlled truck traffic. Preliminary Publication, 7th Congress, IABSE: 497–510.

4. SCHILLING, C. G., K. H. KLIPPSTEIN, J. M. BARSOM & G. T. BLAKE. 1978. Fatigue of Welded Steel Bridge Members Under Variable Amplitude Loadings. NCHRP Report 188. Transportation Research Board. Washington, D.C.

5. FISHER, J. W., K. H. FRANK, M. A. HIRT & B. M. MCNAMEE. 1970. Effect of Weldments on the Fatigue Strength of Steel Beams. NCHRP Report 102. Highway Research Board. Washington, D.C.

FRACTURE CONTROL IN WELDED STEEL BRIDGES

INTRODUCTION

The notion that steel bridges are safe remained unchallenged in the United States until about 1968. The failure of the Point Pleasant Bridge [1] and subsequent problems [2-4] shattered this belief. Terms such as fatigue, fracture, fracture control, fail safe, fracture toughness, and fracture mechanics started to become familiar to those responsible for the design and/or maintenance of steel bridges. While these terms were once viewed as academic and somewhat esoteric, they now represent the building blocks of good safe bridge design. Fracture-safe bridge design and in particular fracture control in welded steel bridges are the main themes of this paper.

Fracture control or fracture safe design for steel bridges can be viewed in simple terms as a series of procedures which provide a structure that will not fail during the design life. The specific definition of failure will vary from structure to structure. A continuously occurring type of cracking that does not impair function and does not lead to personal injury may still be viewed as a failure if the problem requires expensive maintenance programs. Fracture control plans in broad terms try to identify all factors that can contribute to the failure of the structure. In addition, the plans attempt to assess the relative contribution of each factor and evaluate the trade-offs that can be made while at the same time assigning the responsibility for all tasks needed to ensure structural integrity. The body of technical knowledge that provides the information necessary for fracture control is known as Fracture Mechanics. In the remaining sections of this paper a brief discussion of fracture mechanics as it relates to bridge steels will be presented. In addition, five fundamental items that a bridge designer can influence will be highlighted. These are

- Basic structural configuration,
- Specific design detail,
- Operating stress levels,
- Materials of construction,
- Inspection and maintenance programs.

As will be shown these five elements should interact continuously in the design process. A change in one will affect the others.

FRACTURE MECHANICS

Fracture Mechanics as does Strength of Materials, Elasticity, Plasticity and other attempts to describe a material's behavior under load finds its origins related to stress analysis. One of the fundamental observations of Fracture Mechanics was that for a material body which could be represented by a two

Richard Roberts is at the Materials Research Center of Lehigh University, Bethlehem, Pennsylvania 18015.

0077–8923/80/0352–0219 $01.75/1 © 1980, NYAS

dimensional state of stress and where plasticity effects could be ignored the state of stress very near the tip of a crack was uniquely described by the stress intensity factor, K. This takes the form

$$\sigma_x = \frac{K}{\sqrt{2\pi r}} f_x(\theta)$$

$$\sigma_y = \frac{K}{\sqrt{2\pi r}} f_y(\theta) \tag{1}$$

$$\zeta_{xy} = \frac{K}{\sqrt{2\pi r}} f_{xy}(\theta)$$

where r and θ are polar coordinates whose origins are located at the crack tip, K is the stress intensity factor and f_x, f_y, f_{xy} are functions of the coordinate θ. The crack tip element and coordinates are highlighted in FIGURE 1.

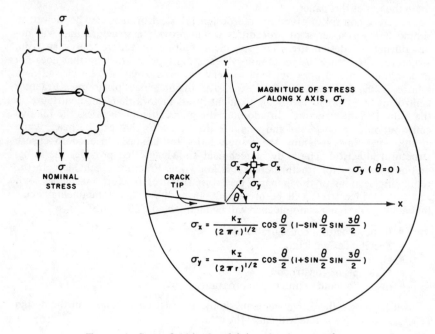

FIGURE 1. Stress field in the vicinity of a sharp crack.

Since this observation was independent of crack configuration, external loading, or geometry, the only way these can affect the stress field is through the quantity K. Thus K reflects the effect of these variables. As such K can be functionally represented as

$$K = f(\sigma, a, W, \ldots,) \tag{2}$$

where f is a function of the applied stress, σ, the crack length, "a", and other geometric dimensions such as plate width, W, etc. For the most part K will take the form

$$K = \sigma\sqrt{\pi a}\, f\!\left(\frac{a}{W}\, ;\, \right) \qquad (3)$$

This shows that K is linearly related to the applied stress and to the square root of crack size times some nondimensional function of crack size to structural dimension W. A more detailed description of the origins of K can be found in references 5 and 6.

For structural situations where the state of stress near the crack tip can be represented by (3) the importance of K is obvious. Since the local stresses at the crack tip are going to control the fatigue and fracture processes in that region, and K is the measure of these stresses, K must be the variable controlling these processes. This leads to the observations that

$$K = \sigma\sqrt{\pi a}\, f\!\left(\frac{a}{W}\right) = K_c \qquad (4)$$

for static loading and

$$\frac{da}{dN} = C\Delta K^n \qquad (5)$$

for fatigue loads. K_c is a critical level of the stress intensity factor; da/dN is the growth of the crack length "a" per cycle where N is the cycle number. C and n are material constants. $\Delta K = K_{max} - K_{min}$ for a cyclic load and the max and min subscripts represent the maximum and minimum levels of K in a cycle.

To understand these relationships consider first (4). What is the difference between K and K_c? This is most easily understood in terms of the simple analogy shown in FIGURE 2. Here a tensile specimen is shown schematically along with the results of a laboratory tensile test. The tensile test results provide data in the form of a load displacement plot from which an estimate of a yield load P_y can be made. This is generally done according to some accepted standard, say ASTM Standard A370.[7] The quantity $\sigma = P/A$ where P is the applied load and A the original specimen area is a man-made quantity. It is purely a man-made mathematical expression that is a result of trying to understand how forces are transmitted through a material body. As a result of a test a quantity P_y can be determined. This gives a corresponding quantity σ_y or a yield strength for the material. $\sigma = P/A$ is the mathematical expectation and $\sigma_y = P_y/A$ is a material property that is the result of a test. The same thing happens in a fracture test. FIGURE 2 shows a compact tension specimen being tested. This consists of loading the specimen which contains a crack of length "a" in a slow controlled manner while simultaneous recording load P and the displacement Δ of the crack surfaces. This also results in a load displacement diagram. Again based on some accepted standard, say ASTM Standard E399,[8] the load and crack length at the onset of fracturing can be determined. This provides data to place in an expression such as (4). In the case of the compact tension specimen

$$K = P\sqrt{\pi a}\, f\!\left(\frac{a}{W}\right) \qquad (6)$$

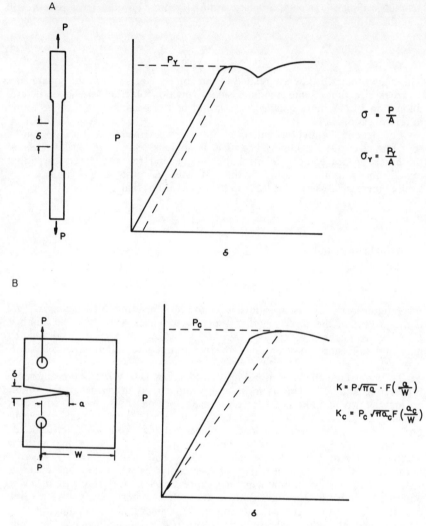

FIGURE 2. (A) Typical tensile test. (B) Typical K_c test.

where the $f(a/W)$ is given in reference 8. As a result of the test there is a set of values P_c, a_c which allows the calculation of a K_c

$$K_c = P_c \sqrt{\pi a_c}\, f\left(\frac{a_c}{W}\right) \tag{7}$$

just as σ_y was calculated. Thus K and K_c represent the mathematical expression and the computed material test result just as σ and σ_y did. In the particular case where the test results meet all of the requirements of E399 the value of K_c is called K_{1c} or the critical plane strain stress intensity factor.

The relationship given by (5) relates the growth of a fatigue crack per cycle to the stress intensity range ΔK and some material constants c and n. For most bridge steels n can be taken as 3. Also ΔK can be represented in many instances as

$$\Delta K = C_1 \Delta\sigma \sqrt{\pi a} \qquad (8)$$

where C is a constant and $\Delta\sigma$ is defined as $\sigma_{max} - \sigma_{min}$. This leads to

$$\frac{da}{dN} = C\, C_1{}^3 \Delta\sigma^3 \pi^{3/2} a^{3/2} \qquad (9)$$

When the final crack length is of the order of ten times the initial crack length this can be integrated to provide an estimate of cyclic life of the form

$$N = \frac{C_2}{[\Delta\sigma^6 a_i]^{1/2}} \qquad (10)$$

where C_2 is a constant and a_i is the initial crack length. The design implications of (10) will be discussed later.

FRACTURE CONTROL IN STEEL BRIDGES

As noted in the introduction there are five fundamental areas which significantly affect fracture control. Some of the key issues related to these will be discussed in this section.

Basic Structural Configuration

The basic structural configuration of the bridge plays a major role in the fracture control plan. A structure with a high degree of redundancy or many parallel load paths will generally tolerate a greater amount of cracking prior to a catastrophic collapse. However, if such cracking occurs in areas that are inaccessible for observation and repair the structure can prove to be seriously inadequate. Major disruptions of traffic due to repairs are simply not acceptable.

The interaction of the basic structural configuration with the other four areas is fairly clear. If the structure is highly crack tolerant because of its basic configuration then more severe measures are not necessary in the other areas.

Specific Design Detail

The specific design details play a major role in fracture control. Improper choice of details and locations can prove to be disastrous. To begin with, all weld details have some degree of stress concentration associated with them. As such each class will give a different fatigue life. This aspect is adequately described by Fisher.[9] However, to appreciate this consider (8) where the $\Delta\sigma$

can be replaced by $K_t \Delta \sigma$. K_t is the theoretical stress concentration. If one chases this through (9) and (10) it is seen that

$$N = \frac{C_2}{[K_t{}^6 \Delta \sigma^6 a_i]^{1/2}} \tag{11}$$

This shows that the number of cycles of life are severely reduced for details which have higher stress concentrations, K_t.

In addition to producing stress concentrations, the specific detail affects the design in terms of workmanship. Some details are more easily inspected and fabricated. This will affect initial detail quality and ultimately the life of the structure. The ease of inspection, repair, and initial fabrication should also affect the choice of weld details.

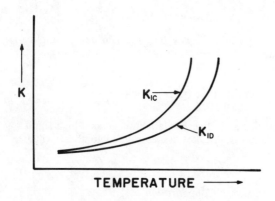

FIGURE 3. Typical dy-, namic, K_{1d}, and static K_{1c} fracture toughness behavior of bridge steels.

Operating Stress Level

Operating stress level probably has more to do with safe bridge design than any other single factor except weld detail. It can be seen in (11) that the number of cycles of life is proportional to the inverse cube of the stress range. Thus, if one wants to make a major improvement in life the two most potent parameters for doing this are K_t and $\Delta \sigma$, the detail configuration and the operating stress level.

Materials of Construction

The choice of the materials of construction has little effect on the fatigue life. It will however interact with K_{1c} in (4). It has been found that K_{1c} for typical bridge steels varies with both loading rate and temperature.[10] This is schematically shown in FIGURE 3. If the stress levels, operating temperature, and loading rates are known for the structure then a required K_{1c} can be specified for the particular steel. It would be desirable to have a level of toughness, K_{1c}, so that for the operating stress σ and the material being used, the critical crack size predicted by **(4)** would be large enough for easy detection. The actual question of what is sufficient toughness is a very difficult one to answer. The American Association of State Highway and Transportation Officials (AASHTO) toughness requirements [11] are an attempt at this. These will be described later.

While material toughness plays a role in determining the final crack size at which a crack will become unstable, it also plays an equally significant role in fabrication. If materials are purchased without any toughness requirements it has been found that this can lead to excessive cracking during fabrication. These cracks can do two things. If they go undiscovered they produce a structure with a reduced fatigue life since the crack initiation stage is reduced. For those cases where the cracks are discovered and repaired the repaired detail is generally not as serviceable as one which has not needed repair. Additional weld repairs increase the probability of service problems. Thus toughness requirements can assist in fracture control by providing a superior initial structure.

Inspection and Maintenance Programs

The initial inspection and maintenance program for a structure will affect the fatigue life. The role of initial crack size a_i compared to K_t or $\Delta\sigma$ is clearly highlighted in **(11)**. To significantly change the fatigue life by changing a_i requires very large reductions in a_i. Similar changes are more easily facilitated by changes of K_t or $\Delta\sigma$.

In addition to the obvious benefits of good inspection and maintenance programs carried out be competent inspectors, a carefully planned inspection and maintenance program for the bridge during the course of its life can provide for the detection of potential failures and allow for timely repairs to be made. Also if the maintenance program is planned during the design stages of the structure, interaction between inspectability and the type and location of the weld detail can be allowed. It is hoped that this will eliminate weld details which are difficult to inspect in service. Also a general plan of accessability can be prepared for all weld details at this time to minimize severe service inspection problems.

AASHTO FRACTURE TOUGHNESS REQUIREMENTS

The current AASHTO toughness requirements [11] are an outgrowth of a proposal by Frank and Galambos.[12] TABLE 1 highlights the current AASHTO requirements. These levels of toughness were not intended to provide any special level of fracture control by themselves. They are intended to provide the bridge designer with a known minimum level of fracture toughness around which a fracture control plan could be constructed. These levels coupled with the choice of design detail, stress level, structural configuration, and the inspection and maintenance program provide the basis for fracture control.

The specific Charpy V-notch (CVN) levels given in TABLE 1 were determined based on the AASHTO design stress levels and typical bridge loading rates. Details of how these were determined can be found in the work of Barsom [13] and Roberts et al.[14] The specific CVN levels were related to K_{1c}

TABLE 1

BASE METAL CHARPY V-NOTCH REQUIREMENTS FOR FRACTURE CRITICAL MEMBERS

ASTM Designation	Thickness, Inches (mm)	Zone 1 *	Zone 2 †	Zone 3 ‡
A36	Up to 4" (101.6)	25 @ 70° F (33.9 Nm @ 21.1° C)	25 @ 40° F (33.9 Nm @ 4.4° C)	25 @ 10° F (33.9 Nm @ −12.2° C)
A572 §	Up to 4" (101.6) mechanically fastened	25 @ 70° F (33.9 Nm @ 21.1° C)	25 @ 40° F (33.9 Nm @ 4.4° C)	25 @ 10° F (33.9 Nm @ −12.2° C)
A572 §	Up to 2" (50.8) welded	25 @ 70° F (33.9 Nm @ 21.1° C)	25 @ 40° F (33.9 Nm @ 4.4° C)	25 @ 10° F (33.9 Nm @ −12.2° C)
A588 §	Up to 4" (101.6) mechanically fastened	25 @ 70° F (33.9 Nm @ 21.1° C)	25 @ 40° F (33.9 Nm @ 4.4° C)	25 @ 10° F (33.9 Nm @ −12.2° C)
A588 §	Up to 2" (50.8) welded	25 @ 70° F (33.9 Nm @ 21.1° C)	25 @ 40° F (33.9 Nm @ 4.4° C)	25 @ 10° F (33.9 Nm @ −12.2° C)
A588 §	Over 2" to 4" (50.8–101.6) welded	30 @ 70° F (40.7 Nm @ 21.1° C)	30 @ 40° F (40.7 Nm @ 4.4° C)	30 @ 10° F (40.7 Nm @ −12.2° C)
A514 ‖	Up to 4" (101.6) mechanically fastened	35 @ 0° F (47.5 Nm @ −17.8° C)	35 @ 0° F (47.5 Nm @ −17.8° C)	35 @ −30° F (47.5 Nm @ −34.4° C)
A514 ‖	Up to 2½" (63.5) welded	35 @ 0° F (47.5 Nm @ −17.8° C)	35 @ 0° F (47.5 Nm @ −17.8° C)	35 @ −30° F (47.5 Nm @ −34.4° C)
	Over 2½" to 4" (63.5–101.6 welded	45 @ 0° F (61 Nm @ −17.8° C)	45 @ 0° F (61 Nm @ −17.8° C)	(Not permitted) (Not permitted)

* Zone 1: Minimum Service Temperature 0° F (−17.8° C) and above.
† Zone 2: Minimum Service Temperature from −1° F to −30° F (−18.3° C to −34.4° C).
‡ Zone 3: Minimum Service Temperature from −31° F to −60° F (−35° C to −51.1° C).
§ If the yield strength of the material exceeds 65 ksi (448.159 MPa), the temperature for the CVN value for acceptability shall be reduced by 15° F (8.3° C) for each increment of 10 ksi (68.947 MPa) above 65 ksi (448.159 MPa). The yield strength is the value given in the certified "Mill Test Report".
‖ ASTM A517 Charpy requirements are the same as for AASHTO M244 (ASTM A514).

values by means of correlations.[13, 14] This was done rather than use K_{1c} testing because the CVN testing is less expensive and is more familiar to most materials people and designers. The particular levels were chosen so that a fairly ductile material behavior could be guaranteed in most failure instances. These levels are also more than sufficient to eliminate the fabrication cracking problems noted previously.

In order to evaluate the AASHTO toughness requirements an extensive research program was carried out at Lehigh University under the auspices of the Federal Highway Administration.[15] In this program a series of 36-in deep beams approximately 24 feet long were fabricated from A36, A588, and A514 material. Four different types of welded details corresponding to various AASHTO fatigue details were fabricated into the beams. These beams were then cycled at the design temperatures and stress levels for two million design cycles. The test results showed that the AASHTO toughness requirements were adequate to guarantee that the beams would sustain at least the design stress cycles at the lowest design temperature. This provided experimental support for the theoretical basis of the AASHTO toughness requirements.

SUMMARY

The problems associated with fracture control in welded steel bridges have received serious consideration in recent years. This paper discusses a number of the key elements that would make up any fracture control plan for a welded structure. The interrelationships among operating stress levels, materials of construction, inspection methods, design details, and basic structural configuration as they affect fracture control are highlighted. In particular the history and rationale for current American Association of State Highway and Transportation Officials material toughness requirements are reviewed and shown to be consistent with experimental evidence.

While many things can and do contribute to fracture control, this paper attempts to show in the strongest of terms that it is ultimately the designers and welding engineers who have the most profound effect on fracture control through the choice of welded details.

REFERENCES

1. Collapse of U.S. 35 Highway Bridge, Point Pleasant, West Virginia, December 15, 1967. National Transportation Report No. NTSB-HAR-71-1.
2. Report of the Royal Commission into Failure of King's Bridge, Victoria, Australia, 1963.
3. Engineering News Record. January 7, 1971.
4. Engineering News Record. March 30, 1972.
5. ROLFE, S. T. & J. M. BARSOM. 1977. Fracture and Fatigue Control in Structures. Prentice Hall.
6. TADA, H., G. R. IRWIN & P. C. PARIS. 1975. The Handbook. Del Research Corporation. Hellertown, PA.
7. ASTM A 370–77 Standard Methods and Definitions for Mechanical Testing of Steel Products. Part 10, ASTM Book of Standards.
8. ASTM E 399–78 Standard Test Method for Plane-Strain Fracture Toughness of Metallic Materials. Part 10, ASTM Book of Standards.

9. FISHER, J. W. 1977. Bridge Fatigue Guide-Design and Details. AISC. New York.
10. ROBERTS, R. 1979. Fracture Design for Structural Steels. *In* Proceedings of 22nd Sagamore Army Materials Research Conference. :165–212. Plenum Press.
11. Guide Specifications for Fracture Critical Non-Redundant Steel Bridge Members, AASHTO, Sept. 1978.
12. FRANK, K. H. & C. F. GALAMBOS. 1972. Application of Fracture Mechanics to Analysis of Bridge Failures. *In* Safety and Reliability of Metal Structures. :279–306. Am. Soc. of Civil Engineers. New York.
13. BARSOM, J. M. 1973. Toughness Criteria for Bridged Steels. Tech. Report No. 5 for AISI Project 168, February 1973.
14. ROBERTS, R., G. R. IRWIN, G. V. KRISHNA & B. T. YEN. 1974. Fracture Toughness of Bridge Steels—Phase 11 Report., Lehigh University, Bethlehem, PA, Dept. of Transportation Contract Report No. FHWA-RD-74–59, September 1974. (PB 239 188)
15. ROBERTS, R., J. W. FISHER, G. R. IRWIN, K. D. BOYER, H. HAUSAMANN, G. V. KRISHNA, V. MORF & R. E. SLOCKBOWER. 1977. Determination of Tolerable Flaw Sizes in Full Size Welded Bridge Details. FHWA Report No. FHWA-RD-77–170.

AERODYNAMIC RESPONSE OF LONG-SPAN BRIDGES TO WIND

FU-KUEI CHANG

I. INTRODUCTION

Since the collapse of the Tacoma-Narrows Bridge on November 7, 1940, and many previous structural failures due to wind, many researchers have studied the causes of the failures both theoretically and experimentally. However, up to the present time (1980), there is no reliable analytical method for predicting the aerodynamic response of bridges to wind action. Recent developments permit the use of wind tunnel tests on models to solve the problem if such tests are carefully planned and include adequate simulation of the actual terrain.

It is the intent of the paper to provide an insight into the problem and to present a state-of-the-art method for predicting the response of bridges to wind. A practical approach used recently in determining the response of an existing major suspension bridge is described.

II. THE NATURE OF WIND AND WIND-INDUCED MOTIONS

The Nature of Wind

Basic Wind Velocity

There are generally two methods for defining the basic wind velocity. One method is to specify the fastest mile wind, which is defined as the average velocity during the time required for the wind to travel one mile. The estimates of the predicted fastest mile at a 30-foot elevation for 2-, 10-, 25-, 50- and 100-year mean recurrence intervals for various localities in the United States are available.[1] Wind velocity estimates for other intervals can also be obtained.[1] The second method is to specify the hourly average wind (or sometimes 10-minute average); i.e., the mean hourly speed.

The fastest-mile method required sensitive velocity measuring instruments, whereas the hourly average is easier to measure and therefore has gained popularity in recent years. The fastest-mile method is still standard in the United States; the relationship between the velocities obtained by these two methods has been established.[2]

Wind Velocities vs. Height

Because of ground friction, wind velocities at any given site vary with height. However, for a certain height above ground, the influence of the ground

---

Fu-Kuei Chang is an Associate with the Consulting Engineering firm Ammann & Whitney, 2 World Trade Center, New York, New York 10048.

0077–8923/80/0352–0229 $01.75/1 © 1980, NYAS

friction on wind becomes negligible, and the velocity remains constant above this elevation. This velocity, V_G, is called the "gradient velocity" and the elevation Z_G, the "gradient height." Davenport [3] has proposed an idealized relationship for different terrains based on power law. Recently, a formulation in terms of logarithmic law has been proposed.[4, 5]

Turbulent Nature of Wind

At a given elevation, the wind velocity is not constant; rather, it varies with time with different intensities for different heights and sites.

The response of structures that are not sensitive to the turbulent wind can be handled by the simplified "gust factor" approach; but for long-span bridges, the structural stiffness, mass and damping cannot be ignored—a different approach will be required.

Channeling and Shielding

The wind velocity at a site is also affected by the presence of adjacent structures and/or special terrain features.

Channeling tends to increase the wind velocity and change its direction, thereby causing unusual and increased pressure and/or suction. On the other hand, direct shielding may result in reduced wind velocity for certain wind directions. These effects can only be determined by properly conducted wind tunnel tests.

Vortex Shedding

Under certain conditions, air flowing past a bridge sheds eddies alternately and creates a periodic force perpendicular to the wind flow; therefore, a bridge might vibrate up and down as well as torsionally, since eddies or vortices shed by the air stream also induce torsional vibration. Resonance might develop between the rate of vortex shedding and some natural frequency of the structure. This is the basic concept of the "vortex shedding theory" in treating the instability of suspension bridges. However, for certain structural shapes at certain wind velocities, the shedding of vortices may become random instead of periodic.

Forces Created by Motions of Structures

The motion of the structure under wind action also creates additional forces. For example, if a bridge is moving downward while a horizontal wind is blowing, the relative wind velocity with respect to the bridge has an upward angle (positive angle of attack). By the same reasoning, upward motion will create a relative wind velocity with a negative angle of attack. If the lift (vertical force) coefficient as measured in static tests shows a slope in the lift vs. angle-of-attack plot that is negative and sufficiently large in absolute value, then it can be shown that the section is susceptible to galloping; i.e., the phenomenon wherein there is a wind force acting downward on the section while the section is moving downward; and upward, while the section is moving upward. This is the basic concept of the so-called "negative slope" theory.

Flutter

The classical flutter theory as applied to streamlined objects such as aircraft wings involves the coupling of vertical and torsional modes. For bridges that are not streamlined, this type of flutter had never occurred. However, it is possible for bridges to vibrate violently in single-degree-of-freedom (torsional) mode. In the flutter-type motion, the higher the wind velocity, the more violent the motion becomes. In other words, the aerodynamic forces tend to reduce the overall (structural and aerodynamic) damping of the structure and eventually make it negative, and therefore the structure will enter a diverging vibration condition. Sometimes, this is also called the "negative damping" theory.

There is no analytical solution to the flutter problem. Scanlan [6] proposed to attack the problem by experimentally determining various flutter derivatives (or coefficients) in laminar flow so that the forcing functions can be formulated.

Wind-induced Motions

As described previously, the wind force consists of a steady component or static force, and several time-dependent components or dynamic forces. The motions induced by these forces and certain properties of the structure which affect the motions are given in TABLE 1.

From various model tests conducted mainly in the United States,[7] Switzerland [8] and England,[9] the wind pressure coefficients over various surfaces and bridge shapes are available. From these coefficients, the static or quasi-static component of the wind force can be determined.

Once the force is known, it is a simple matter to find the response and deflection. However, to evaluate the response of a bridge to dynamic components of the wind is a very complex problem. Indeed, it is very difficult to formulate mathematically the forcing functions for a given site. The forces induced by motions of the structure are also unknown. The structural dynamic properties, especially damping, are also difficult to determine analytically. Not knowing the forcing functions, the forces induced by motions of the structure and damping, it is virtually impossible to evaluate the response of the structure. It is then rational to approach the problem whenever possible as follows:

- Find the dynamic structural properties by ambient vibration survey. This will be discussed in detail in Section III.
- Determine the forcing functions and response by properly planned wind tunnel investigation, monitor the bridge motion by full scale instrumentation, if possible, and modify the wind tunnel predictions, if warranted (Sections IV).
- Evaluate the effects of the response (Section V).

TABLE 1

BRIDGE RESPONSE TO WIND

Wind	Structural Property	Response	Effect
Static Force	(Shape) Stiffness	Deflection	Stability Strength
Dynamic Force	(Shape) Stiffness Mass Damping } Frequency	Vibration	Stability Strength Users' Response

III. DETERMINATION OF DYNAMIC STRUCTURAL PROPERTIES

Dynamic Structural Properties

As can be seen from TABLE 1, the structural properties that affect the response of dynamic forces are frequency and damping. Although it is possible to compute mathematically the natural frequencies for various vibration modes, their accuracies are questionable at times because of simplified assumptions used in the analysis. As for damping, there is no method to compute its value. Fortunately, all these dynamic properties can be measured in the field by the use of the Ambient Vibration Survey (AVS).

Ambient Vibration Survey

In the Ambient Vibration Survey (AVS), the response of the structure to motion caused by traffic, wind, etc., is measured with sensitive seismometers. The recorded motion is analyzed using the spectral techniques. The direct power spectral density of each recorded motion yields estimates of natural frequencies and modal damping; whereas a cross-spectral density between a reference record and all other records yields mode shape estimates.[10]

In the AVS of a major suspension bridge, six force-balance accelerometers were used in the field operations. The accelerometers were arranged in six different patterns or set-ups for testing the bridge. One accelerometer was at a certain panel point during each set-up. This served as a reference for normalizing all the measurements. A calibration run with all accelerometers side by side was made at the beginning of the survey. The entire AVS analysis procedure is a statistical process and is dependent on the signal-to-noise ratio in each mode. Where the signal is large, the statistical reliability of the mode value is high and vice versa. The procedure used in the AVS for determining the damping is called the "bandwith method" or "Q-factor method," which gives the damping in terms of percent of critical. Critical damping is the minimum amount of damping that will eliminate vibration. All the data reduction is done by computer on the basis of the work of Crawford and Ward.[11] The results obtained are summarized in TABLE 2 and FIGURES 1 and 2. TABLES 3 and 4 give the measured natural frequencies and damping coefficients for two other bridges.

The damping data mentioned above were all obtained under low vibration amplitudes. There are indications that the damping will increase for larger amplitude motions. However, the yielding tends to soften the stiffness and therefore increase the period. As far as dynamic response is concerned, these two effects more or less compensate each other.

One method of measuring the damping at large amplitude is to monitor the motions of the structure by actual instrumentation. The damping can be derived by digitizing the motion records and performing a spectral analysis. However, for large amplitude motions, the aerodynamic damping will be appreciable and must be deducted from the derived damping in order to obtain the proper damping value (mechanical damping). Accurate determination of the aerodynamic damping is extremely difficult. It is usually on the conservative side to assume a constant damping measured from the Ambient Vibration Survey under moderate wind conditions. For the William Preston Lane Memorial Suspension Bridge in Maryland, the ambient vibrations were recorded with both strong wind blowing and with practically no wind. There was no noticeable difference in the damping measured in each case.[10] However, for this bridge, there is no large motion of any kind under strong wind. For bridges excited into major motions, there might be an increase in damping.

TABLE 2

SUMMARY OF MODES OF VIBRATION, LONG-SPAN SUSPENSION BRIDGE

Mode	Period (sec)	Damping (% of critical)	Classification
1	8.77	2.9	First Lateral
2	5.35	1.3	First Symmetric Vertical
3	4.97	1.9	First Anti-Symmetric Vertical
4	3.65	1.11	First Symmetric Torsional
5	3.33	0.9	Second Symmetric Vertical
6	3.11	3.18	Second Lateral
7	2.85	0.85	Second Anti-Symmetric Vertical (Large on Side Spans)
8	2.76	0.85	First Anti-Symmetric Torsional
9	2.44	0.95	Third Anti-Symmetric Vertical (Large on Side Spans)
10	2.37	0.81	Second Symmetric Torsional
11	2.31	0.76	Third Anti-Symmetric Vertcial
12	1.95	0.53	Second Anti-Symmetric Torsional (Large on Side Spans)
13	1.79	0.52	Fourth Symmetric Vertical
14	1.75	0.46	Third Symmetric Torsional (Large on Side Spans)
15	1.68	0.50	Third Anti-Symmetric Torsional
16	1.42	0.66	Fourth Anti-Symmetric Vertical
17	1.35	0.50	Third Lateral
18	1.29	0.76	Fourth Symmetric Torsional
19	1.13	0.71	Fifth Symmetric Vertical
20	1.02	0.50	Fourth Anti-Symmetric Torsional

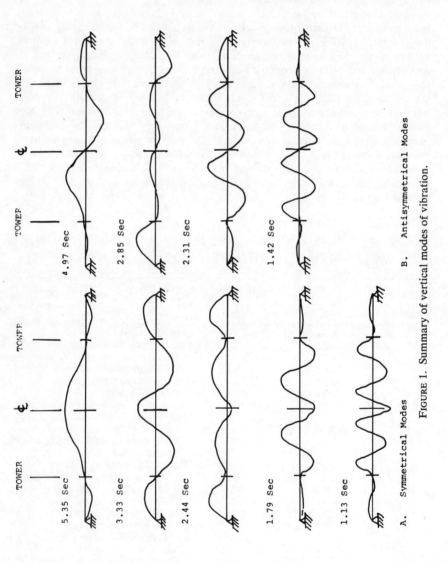

FIGURE 1. Summary of vertical modes of vibration.

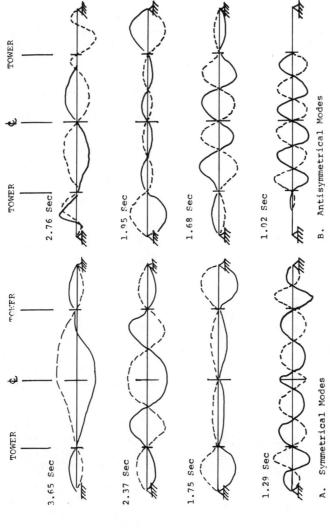

A. Symmetrical Modes

B. Antisymmetrical Modes

FIGURE 2. Summary of torsional modes of vibration.

TABLE 3

SUMMARY OF MODES OF VIBRATION, NEWPORT BRIDGE

Mode	Period (sec)	Damping (% of critical)	Classification
1	6.45	3.0	First Lateral
2	5.42	2.0	First Symmetric Vertical
3	4.53	1.7	First Anti-Symmetric Vertical
4	3.2	2.0	Second Anti-Symmetric Vertical
5	3.00	1.1	Second Symmetric Vertical
6	2.54	1.1	Third Symmetric Vertical
7	2.38	1.1	Second Lateral
8	2.20	0.8	First Symmetrical Torsional
9	1.92	1.2	Third Lateral
10	1.85	0.9	Third Anti-Symmetric Vertical
11	1.62	0.6	First Longitudinal Tower
12	1.56	0.5	Fourth Lateral
13	1.54	0.8	First Anti-Symmetric Torsional
14	1.45	0.6	Fourth Symmetric Vertical (Side Spans)
15	1.42	0.6	Fifth Lateral
16	1.34	0.5	Second Longitudinal Tower
17	1.32	0.8	Fifth Symmetric Vertical
18	1.24	0.8	Third Longitudinal Tower
19	1.22	0.9	Sixth Lateral
20	1.006	0.4	Fourth Anti-Symmetric Vertical

TABLE 4

SUMMARY OF MODES OF VIBRATION, CHESAPEAKE BAY BRIDGE

Mode	Period (sec)	Damping (% of critical)	Classification
1	9.52	5.4	First Lateral
2	4.88	2.5	First Symmetric Vertical
3	3.85	2.2	First Anti-Symmetric Vertical
4	3.13	1.4	Second Lateral
5	2.86	1.4	Second Anti-Symmetric Vertical
6	2.41	1.1	Second Symmetric Vertical
7	2.08	1.6	Third Symmetric Vertical
8	1.71	0.5	Third Lateral
9	1.59	1.3	Fourth Lateral
10	1.481	1.4	First Symmetric Torsional
11	1.351	0.9	Third Anti-Symmetric Vertical
12	1.16	0.9	Fifth Lateral
13	1.07	0.4	Sixth Lateral

In the analysis and design of a new bridge, many available analytical methods and computer programs can be used in determining its natural frequencies and mode shapes. However, its damping coefficients can only be evaluated by judgment, taking into account the measured values of similar structures.

IV. Determination of Structural Response

Response to Static Forces and in the Absence of Dynamic Effects

It can be seen from Table 1 that once the static forces are known, the response of a bridge to these forces can be found if the stiffness is known; therefore, the determination of response to this type of loading amounts to the determination of the effective wind forces. However, for long-span bridges in unusual terrains, special studies on velocity pressure coefficients, gust factor and net pressure coefficient are required.

Response to Dynamic Forces

It is virtually impossible to predict analytically the response of wind-sensitive structures in different terrains to dynamic forces caused by turbulence, vortex shedding, forces created by motions of structures, flutter, etc., since these forces are generally not mathematically determinable. The only approach is to evaluate the response by wind tunnel investigation.

There are two classical questions regarding the validity of wind tunnel testings:

- What is the scaling factor or Reynolds number effect?
- What is the wind tunnel wind vs. natural wind effect at site?

In answering the first question, it appears that for bridges with blunt edges, under turbulent winds the scaling parameter related to a fluid dynamic similarity criterion is generally acceptable. However, the second question presents a real problem in any wind tunnel investigation.

The simulation of the natural wind at a specific location in the wind tunnel investigation involves the analysis of all available full-scale wind data, the examination of the structure surroundings and, if necessary, the testing of a topographic model. Wind from different azimuth directions may have different characteristics, which can be simulated in the aeroelastic model tests by introducing different surface roughnesses so that a boundary layer turbulent flow similar to the natural wind at site for various azimuth directions can be obtained. In the analysis of the test results, consideration should be given to the fact that the strong winds are not always from the critical response direction. The probability distribution of wind speed for a specific area may be determined by analyzing statistically the available strong wind data including hurricane records.

As for the "angle of attack", it can be assumed that the wind always blows in the horizontal direction. Actually, because of the topographic features of the terrain, horizontal wind can generate forces with different "angles of attack" at different parts of a structure in the wind tunnels as well as in the actual surroundings. Recent measurements [12] indicate that during a hurricane, local forces with angles of attack up to ±15 degrees are possible. However, the wind speed decreases rapidly with the increase of angles of attack.

Wind Tunnel Testing of Suspension Bridge Models

In the early stage of wind tunnel investigation of suspension bridges, section models were used. However, the turbulent boundary-layer nature of the wind, the three-dimensional effect of the full-length bridge and the various wind azimuth directions cannot be simulated in the tests. Davenport[13] suggested the use of aeroelastic models to overcome all the shortcomings of the section model tests. To do this, a rig was developed in which sections of the deck, whose mass distribution and inertia were correctly scaled, were suspended on pairs of parallel, taut piano wires running at the level of the shear center. The decks thus vibrated as taut springs primarily in a fundamental half-wave mode shape. By adjusting the tension and wire spacing, the correct ratio between torsion and vertical motion could be obtained. These models have been termed "taut strip" models.

In the wind tunnel investigation of a major suspension bridge, a 1:60 section model and a 1:400 taut strip model were used in the tests. The section models were tested in a uniform smooth flow and a uniform turbulent flow. The taut strip models tested in a turbulent boundary layer flow were representative of natural wind at the full scale site for various azimuth directions.

The results of both the section model and taut strip model tests indicate that the dominant dynamic response is torsional. Examination of the test results indicates that the dependence of responses upon wind velocity in taut model tests were more gradual than in the case of the section model tests. Unless extrapolated analytically by taking into account the natural mode of vibration of the bridge, the section model in uniform smooth wind flow tends to give predictions that are too conservative. The taut strip model tests, which consider the full length of the bridge as well as the natural wind at the bridge site with simulated turbulent boundary layer flow, are believed to provide more accurate estimates of the full-scale bridge behavior under high winds.

The predicted bridge motions for various combinations of wind speed and wind direction, based on the taut strip model tests, are given in FIGURE 3. The gradient wind speed in FIGURE 3 is defined as the wind speed at gradient height of 1,800 ft. To obtain the wind speed at the deck level, the gradient wind speed should be multiplied by a factor of 0.70. For wind speed at a 30-ft elevation, a factor of 0.54 should be used. The horizontal scale at top of FIGURE 3 gives the wind speed at the deck level. The total single amplitude peak deflection in FIGURE 3 is defined as the maximum vertical deflection of the stiffening trusses, either upward or downward, due to combined vertical and torsional motions.

Full Scale Instrumentation for Dynamic Response

For this bridge, permanent displacement measuring instrumentation was installed in 1971. The records of the action motions of the bridge for several wind storms are available for comparison with the predicted values.

The instrumentation consists of four sensors attached to the center span of

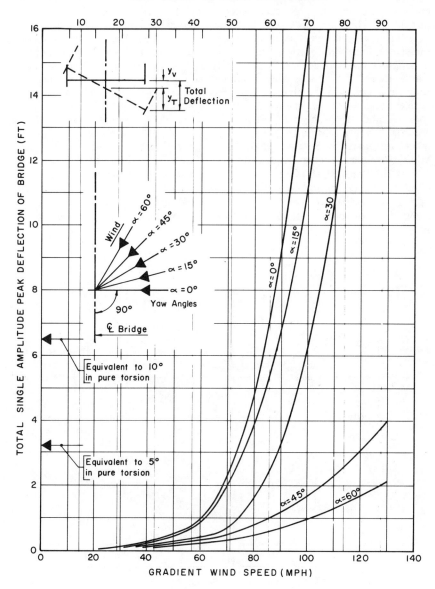

FIGURE 3. Predicted dynamic deflection.

the bridge and a recording apparatus in the Administration Building. Recording is started when an anemometer on its recorder shows the strong winds are blowing. The anemometer and its recorder were installed a few years earlier. FIGURE 4 shows the location of the sensors, which are designated as A, B, C and D. The sensors are accelerometers which operate on a force balance principle. When an acceleration displaces an internal pendulum, it is driven back to its original position by the sensor mechanism. The output voltage of the sensor unit is proportional to the force necessary to hold the pendulum in its original position and is thus proportional to acceleration. This signal is transmitted to the console which contains the logical circuitry that converts the four sensor acceleration signals to a paper recording of the motion. It converts accelerations to displacements through double integration, combines them as specified and drives the pen and paper. FIGURE 5 is a sample of the motion records during a 54-mph storm. Traces 1, 2, 3 and 4 are the direct displacements measured by Sensors A, B, C and D, respectively. One-half the sum of Displacement A plus Displacement B (Trace 6) and one-half the sum of C plus D (Trace 8) show the vertical components of the motion of the center divider at the center and quarter points, respectively. One-half the displacement at A minus the displacement at B (Trace 5) and one-half C minus D (Trace 7) show the torsional rotation of the deck at the center and quarter points, respectively.

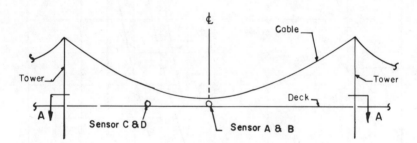

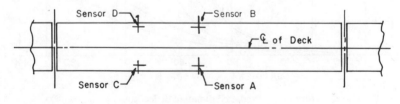

PLAN A-A

FIGURE 4. Sensor locations.

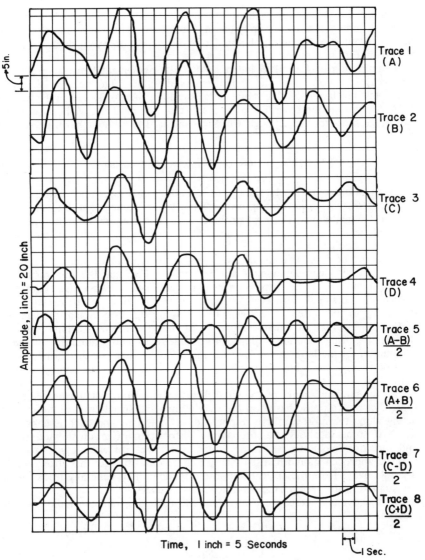

FIGURE 5. Recorder output.

Measured vs. Predicted Dynamic Response

The records of the actual motions of the bridge during wind storms for the period from November 13, 1971 to November 8, 1972 were examined. Nearly all strong winds on the bridge during this period were from east to west, with a yaw angle "α" of 29 degrees. The yaw angle is defined as the angle between the wind and the normal to the centerline of the bridge. The actual measured motions of the bridge during these storms, modified to reflect the deflections of the stiffening trusses, are plotted in FIGURE 6. The curve for predicted motion is obtained by interpolation from the curves for yaw angles of 15° and 30° in FIGURE 3.

Examination of FIGURE 6 indicates that the agreement between predicted and measured motions in the range observed, taking into account that the directional resolution of the actual data is ±11¼°, is fairly good. However, the maximum motion recorded so far is less than the predicted value. Actually, the bridge has experienced high winds during its life without harmful effects.

V. EVALUATION OF DYNAMIC RESPONSE

As mentioned previously, the dynamic motions of bridges under wind action can be predicted by performing proper wind tunnel testings and by using full-scale instrumentation. The next problem is to determine whether the motions of the structure are acceptable or not. This can be a more difficult task than the motion prediction itself.

It should be noted that the dynamic motions may change from one mode to another, or to a combination of several modes. For example, the measured dominant motion of the suspension bridge mentioned previously is symmetric vertical and torsional. However, if the center ties which connect the centers of the cables to the stiffening trusses for restraining the longitudinal movements of the cables break or become loose, as in the case of the Tacoma-Narrows Bridge, the bridge motion will probably change to anti-symmetric torsional. The motion of the latter mode is much more violent than for the former mode because there is no cable length change and, consequently, the energy absorption is low.

As for the yaw angles of the wind direction, the conventional procedure is to predict the response based on the maximum wind speed from any direction acting along the most sensitive response direction. This approach is conservative. At some localities, the probability distribution of gradient wind speed and direction including hurricane winds is known. In such cases, a more realistic approach should be used taking into account the fact that strong winds are not always from the critical response direction. This generally leads to less severe motions for most of the wind storm occurrences.

In the dynamic response evaluation, the following effects should be considered.

Stability

The section model tests of suspension bridges performed in the 1950s indicated that at a certain wind velocity called the "critical wind velocity," some bridge decks will become aerodynamically unstable. In other words, beyond the critical velocity, deck deflections will increase significantly for a very small increase in wind velocity.

The results of the section model tests using smaller damping factors suggest this danger. However, in the taut strip model tests simulating the full length of the bridge and the natural wind at the site, the slopes of the peak deflection vs. wind speed curves, as shown in FIGURE 3, are not too steep in the range of

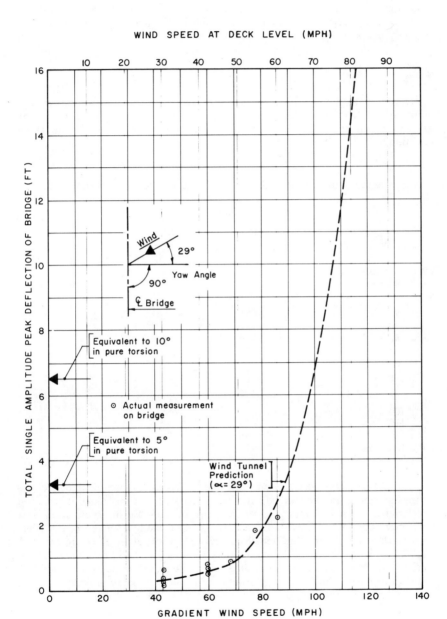

FIGURE 6. Predicted dynamic deflection vs actual measurement ($\alpha = 29°$).

anticipated wind speeds. In other words, with the increase in wind speeds, the structure simply deflects a little more. There is no critical wind velocity as such. In these cases, it is no longer a problem of stability, but rather one of strength that is involved.

Strength

During dynamic motions, accelerations will be generated at the various points of a structure. The accelerations generated as a percentage of the gravity, g, due to peak deflections can be computed by using the following equation [14]:

$$\text{Fraction of } g = 0.1023 \ Af^2$$

where A is the amplitude in inches and f is the frequency in hertz.

TABLE 5

DEFLECTION VS. ACCELERATION

Single Amplitude Peak Deflections (in)	Equivalent Angular Single Amplitude Twist	Maximum Added or Reduced Acceleration at Mid-point of Center Span (%)
20	2°-35'	15
30	3°-52'	23
40	5°-09'	31
50	6°-26'	38
60	7°-41'	46
70	8°-58'	54
80	10°-13'	61
90	11°-27'	69
100	12°-42'	76
110	13°-55'	84
120	15°-08'	92
130	16°-19'	100

TABLE 5 gives the acceleration generated by 20- to 130-inch peak deflections of the suspension bridge studied. It should be noted that the added or reduced acceleration given in TABLE 5 is the increase or decrease in acceleration over and above the normal gravity acceleration g. For example, if the peak deflection reaches 130 inches, the added or reduced acceleration given in TABLE 5 is 100% g; with added 100% acceleration, the dead load will be doubled. This might be a strength problem depending on the factors of safety used in designing various parts of the bridge. During the 100% g acceleration reduction with "zero g," any object that is not properly anchored will float.

VI. SUMMATION AND CONCLUSIONS

The wind force consists of two components: mean and fluctuating forces. The response of a bridge to the former force is now well understood. However, the response to the latter force is a very complex phenomenon that still requires research.

There are various theories that explain the causes for the dynamic motions of bridges under wind action. However, none of the theories can predict quantitatively the nature and magnitudes of the motions.

For long-span bridges, which are sensitive to wind action, a rational approach to the problem is to proceed whenever possible as follows:

(1) Measure vibration modes, frequencies and damping by performing an ambient vibration survey.

This survey is to be used for existing structures only. For the design of new structures, vibration modes and frequencies can be computed analytically. However, the damping coefficient has to be evaluated by judgment, taking into account the measured values for similar structures.

(2) Perform properly planned wind tunnel testing—the entire structure as well as the natural wind at the site should be simulated as carefully as possible.

(3) If possible, monitor the dynamic motions by full-scale instrumentation. Modify the wind tunnel predictions, if warranted. The changes in damping and frequencies for large magnitude motions can also be verified.

The predicted dynamic motions are generally given in terms of peak response vs. return period of the wind (in years). The designer, the owner and/or the appropriate regulatory agency has to decide (a) the life expectancy of the structure and (b) an acceptable degree of risk.

(4) The effects of the motions on the structure as a whole as well as on each component should be carefully evaluated. The effects on users' comfort should also be considered.

SUMMARY

The problem of wind effects on bridges has long been one of great interest to the engineering profession as well as to the general public. This paper deals with various aspects of wind loading and the corresponding response of bridges with reference to model testing and actual motion measurements. Conclusions and recommendations contain a state-of-the-art report on the subject.

ACKNOWLEDGMENT

FIGURE 3 is based on a private report by Drs. A. G. Davenport, N. Isyumov and H. Tanaka of the University of Western Ontario.

REFERENCES

1. THOM, H. C. S. 1968. New Distributions of Extreme Winds in the United States. Journal of the Structural Division, ASCE, July.
2. DURST, C. S. 1960. Wind Speeds Over Short Period of Time. Meteor Magazine, Vol. 89.

3. DAVENPORT, A. G. 1961. Rationale for Determining Design Wind Velocities, ASCE Transactions, Part II.
4. SIMIU, E. 1973. Logarithmic Profiles and Design Wind Speed. Journal of the Engineering Mechanics Division, ASCE, October.
5. INTERNATIONAL STANDARDS ORGANIZATION. 1977. Draft ISO Standards on Wind Load and Wind Design of Structures. Working Document, ISO-TC 98/SC 3/WG2, TNO Construction Research Institute, Delft, Holland, April 1977.
6. SCANLAN, R. H. 1975. Recent Method in the Application of Test Results to the Wind Design of Long, Suspended Span Bridges. FHWA Report No. R.D.-75-115, October.
7. ASCE TASK COMMITTEE ON WIND FORCES. 1962. Wind Forces on Structures. ASCE Transactions, Part II.
8. Schweizerischer Ingenieur und Architekten Verein. Technische Normen No. 160. 1956.
9. SCRUTON, C., & C. W. NEWBERRY. 1963. On the Estimation of Wind Loads for Buildings and Structural Design. Proceedings of the Institution of Civil Engineers, Vol. 25, June.
10. McLAMORE, V. R., G. C. HART & I. R. STUBBS. 1971. Ambient Vibration of Two Suspension Bridges. Journal of the Structural Division, ASCE, October.
11. CRAWFORD, R., & H. S. WARD. 1954. Determination of the Natural Periods of Buildings. Bulletin of the Seismological Society of America. Vol. 54, December.
12. GADE, R. H., H. R. BOSCH & W. PODOLNY, JR. 1976. Recent Aerodynamic Studies of Long Span Bridges. Journal of the Structural Division, ASCE, July.
13. DAVENPORT, A. G., N. ISYUMOV & T. MIYATA. 1971. The Experimental Determination of the Response of Suspension Bridges to Turbulent Wind. Third International Conference on Wind Effects on Buildings and Structures, Tokyo, Japan.
14. CHANG, F. K. 1973. Human Response to Motions in Tall Buildings. Journal of the Structural Division, ASCE, June.

CATASTROPHIC AND ANNOYING RESPONSES OF LONG-SPAN BRIDGES TO WIND ACTION

Robert H. Scanlan and Joseph W. Vellozzi

Introduction

It is well known that even very large and heavy bridges occasionally respond to wind in undesirable ways. That some of the undesirable types of response may even "get out of hand" and lead to catastrophe is therefore a lurking fear. The main problems of long-span bridges under wind may be grouped under the headings of vortex-induced motion, flutter instability, and wind buffeting. In the present paper, each of these types of excitation will be discussed briefly with the intent of emphasizing those aspects that are of interest to the designer. In particular, the elements of design that cause the problems will be mentioned, and the kinds of modifications that alleviate them will be discussed.

Vortex Shedding

In the classic case of vortex shedding, a long circular cylinder, when placed in a cross flow, sheds vortices alternately into its wake, forming a so-called "Karman vortex trail." The same phenomenon is also readily reproduced when the long cylinder is not necessarily circular in cross section, but has any bluff shape like square, rectangular, or otherwise. The vortices in all cases are shed according to the Strouhal rule

$$\frac{fD}{V} = S$$

where f is shedding frequency in Hz, (a pair of alternate vortices per cycle), D is the projected across-wind dimension, V is wind velocity, and S is the Strouhal Number, which ranges from 0.11 to 0.25, depending mainly on the shape of the bluff object. For a circular cylinder, $S \cong 0.2$; and for a square, $S \cong 0.12$.

Bridge decks are in general bluff objects and are thus among the group that can shed vortices. For example, the cross section of the original Tacoma Narrows bridge, sketched in Figure 1, shed vortices at a value of S estimated at 0.16 to 0.18.[1, 2] The original cross section of the Long's Creek bridge, studied by Wardlaw,[3] is shown in Figure 2. This short, cable-stayed bridge shed vortices strongly enough to cause center-span vibration amplitudes to reach about 1 foot.

The two basic aerodynamic cures for vortex shedding in the context of bridge decks are streamlining on the one hand and the manipulation of

Robert H. Scanlan is in the Department of Civil and Geological Engineering of Princeton University, Princeton, New Jersey 08540. Joseph W. Vellozzi is with the engineering firm Ammann and Whitney, 2 World Trade Center, New York, New York 10048.

0077-8923/80/0352-0247 $1.75/1 © 1980, NYAS

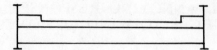

FIGURE 1. Rough outline, original Tacoma Narrows Bridge —cross-section.

"porosity" or through-flow effects on the other. FIGURE 2 also shows how the bluff, solid section of the Long's Creek bridge was streamlined, in the presence of a water surface, to quiet vortex-induced effects. FIGURE 3 depicts that actual bridge with fairings in place.

Through-flow, vortex-suppression effects are often achieved incidentally to the design form for truss-stiffened, long-span bridges, the open gridwork of the trusses acting like vortex "shredders" which prevent coherent vortices being shed in organized fashion. By and large, open-truss bridge decks are not severely subject to the vortex-shedding phenomenon.

Vortex excitation for whole decks is in general a low-windspeed phenomenon when it does occur. It is only of engineering interest when f, the vortex-shedding frequency, is very near a natural structural frequency for across-wind motion.

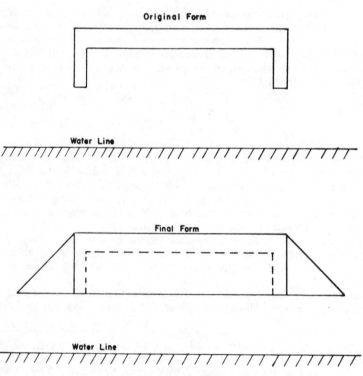

FIGURE 2. Basic form of Long's Creek Bridge deck and modifications.

It should be mentioned here that vortex excitation can be a very annoying problem for individual structural members, like wide-flange or I-sections at fairly high wind velocities and $S = 0.12$ to 0.14. It also can be a problem during erection, for free-standing towers, as in the case of the newer Forth bridge, where preventive friction dampers had to be introduced via long cables to quiet the excessive motion.

The main cures (cf. ref. 19) for vortex-induced response of individual structural members have been through opening holes in the beam webs, stiffening the members greatly by adding material that boxes them in, installing aero-

FIGURE 3. Long's Creek Bridge.

dynamic spoilers on their surfaces (usually an unsightly fix) and by the use of tuned-mass dampers (TMDs), as was the case on the Commodore Barry Bridge across the Delaware at Chester, Pennsylvania. The action of TMDs is to suppress the "lock-on" of vortex shedding to structural motion by inhibiting the initial motion itself. FIGURE 4 suggests a TMD mounted on a structural member.

The lock-on phenomenon is actually the most important aspect of vortex-associated oscillations. In short, vortices are naturally shed by the wind from a fixed structure, but if their shedding frequency excites a structural frequency, the structure displaces rhythmically and begins to control the vortex production. This latter, locked-on to the structural movement, does not then follow the simple Strouhal-prescribed rhythm but holds, for a range of wind velocity, to

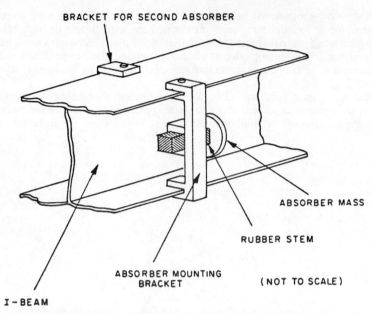

BRACKET FOR SECOND ABSORBER

ABSORBER MASS

RUBBER STEM

ABSORBER MOUNTING
BRACKET

(NOT TO SCALE)

I-BEAM

FIGURE 4. Conceptual form of TMD on structural member.

the structural rhythm. This phenomenon can be merely annoying, as in the Long's Creek Bridge case, or it could lead to serious fatigue of a structural member, as did occur in the case of certain members of the Barry Bridge structure.

The study of bridge deck section models in the wind tunnel is the surest way to detect their vortex-shedding proclivities. For example, the studies of Gade and Bosch [4] on section models of the Luling Bridge are good examples of preoccupation with vortex suppression through streamlining.

The original Tacoma Narrows deck was subject to vortex shedding, most of the action of which led to vertical bending oscillations of that bridge. Simple vortex shedding, however, was not the cause of the final motion and collapse of that structure, as will be discussed subsequently.

It should be mentioned that numerous cases of vortex-induced oscillation are associated with squat H-sections typified by such girder-stiffened decks as those on the original Tacoma Narrows, Deer Isle, and Bronx-Whitestone bridges. A case of vortex-induced oscillation was observed during erection on one of the temporarily unsupported cantilevered extensions of an approach roadway to the Dartmouth-Halifax Bridge in Nova Scotia. It will occur every chance it gets. The ingredients are simply: a bluff body, with freedom to move elastically, and exposure to a steady wind. These are the danger signs for the designer.

FLUTTER

The Tacoma Narrows experience of course stands as the classic example that, aerodynamically, things can in fact get out of hand and lead to catastrophe.

The flutter phenomenon of bridges is now much better understood than at any time in the past. Here again the solution approaches are twofold—either create aerodynamically "porous" structures or streamlined ones.

The truss-stiffened structures have not generally been found stable unless porosity was extended to the deck surface, where gridworks, allowing through-flow from below to above, were included. Such solutions were first suggested by Egli and Ackeret, in Switzerland, and were adopted by Steinman in his Mackinac and Tagus River bridges. Recently such approaches have been included in Japan in the Honshu-Shikoku projects. FIGURE 5 depicts a cross section of that project.

Streamlined bridges were pioneered in Britain, where they came into being when a box structure for the Severn project, long under wind tunnel study by Scruton, was streamlined at the National Physical Laboratory and became a prototype for streamlined bridge deck sections. Later the Lillebaelt was also streamlined through studies by Selberg and Hjorth-Hansen.[5] Wardlaw has used a somewhat different streamlining approach, earlier on a project for the Burrard Inlet,[6] with which Leonhardt was associated, and more recently on the Pasco-Kennewick[7] in the state of Washington, designed by Grant and Leonhardt. FIGURE 6 shows cross-sections of the Severn, Lillebaelt, Burrard Inlet project, Pasco-Kennewick, and Humber designs.

Flutter is an unstable oscillation that grows in amplitude when the right cross-wind velocity (usually rather high) is reached. There are several kinds of flutter, of which the two most important for bridges are *classical* type and *single-degree torsional* type.

Classical flutter was first seen in aircraft, and it involves both bending and torsion motions of the airfoil, coupled together, with a definite phasing between the motions. This two-degree-of-freedom flutter type can also theoretically take place for streamlined bridge decks, but in fact it usually requires such a high wind velocity that it cannot practically take place within the windspeeds normally seen at bridges (say up to 125 mph). This is the main reason why streamlined bridges have a desirable form.

On the other hand, unstreamlined bridge decks (truss-or girder-stiffened) are all strongly subject to single-degree-of-freedom torsional flutter. This is the kind of potential instability that the bridge designer must be most acutely aware of. This type of flutter, related to the stall flutter of airfoils, is associated with

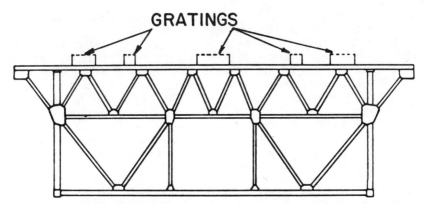

FIGURE 5. Modern open-truss deck section (Honshu-Shikoku Project).

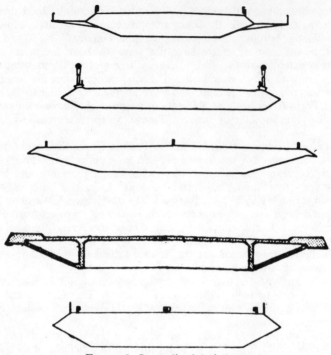

FIGURE 6. Streamlined deck forms.

flow separation from the side and top surfaces that may be struck by a cross wind.

The main mechanism of all types of flutter is that it is *self-excited;* that is, once a structural motion is set up by any cause, that motion causes wind pressures of a destabilizing nature to build up and enhance the motion. The main element in the destabilization appears to be the separation of the flow from the leading corners of the bridge cross-section as the section twists. This explains the thread of similarity that runs through all the designs depicted in FIGURE 6, where the leading edge is so shaped in all cases as to reduce flow separation when the deck section rotates.

This flow-separation phenomenon is particularly acute when the structure section is bluff and non-porous, as in the case of the original Tacoma Narrows, the Deer Isle, the Bronx-Whitestone, and the more recent Sitka Harbor bridge.

When a section like one of these is not moving it may trip vortices above and below the deck as suggested in FIGURE 7. On the other hand, when the section is rotated, the vortex pattern is distorted, as also indicated in FIGURE 7, and the vortex remaining nearest to the deck creates a high suction tending to further twist the deck. At some critical velocity, the vortices travel across the deck in a rhythm of destabilizing action that coincides with a natural torsional frequency and leads, if unchecked, to catastrophic flutter.

This was, in fact, the mechanism that destroyed the original Tacoma Narrows bridge, in a catastrophic fundamental torsion mode that was not seen at all, according to Farquharson,[2] prior to about one hour before the collapse. It was this flutter mode and mechanism—and not simple vortex shedding in the usual sense—that did the ultimate damage at Tacoma Narrows.

Wind Tunnel Models in Bridge Aerodynamics

Very early on, in 1940, even before the Tacoma Narrows collapse, wind tunnel models of that bridge were being tested, notably by Farquharson.[2] Full-bridge models being costly, the section model was chosen early as the most efficient device for study purposes. This consisted of a rigid model of a typical deck section mounted on springs to simulate the bending and twisting of the full-scale structure. Plates were usually installed parallel to the flow at either end of the model to insure that there would be no falsification of the flow pattern at the ends of the model span. Figures 8 and 9 [20] depict typical section models set up for wind tunnel study.

Although long used and conceived of as a typical dynamic analog of the full bridge, the section model is most properly regarded as first, a device for assessing and adjusting the proper aerodynamic contour of a bridge deck; and secondly, as an analog computer of the motion-dependent aerodynamic forces that act on the deck section. The first of these test models is the one that is most particularly valuable during the design stage. Later, when the contour has been fixed, experimental determination of the so-called "flutter derivatives" (see ref. 8) provides the necessary background for a variety of calculations concerned with expected bridge performance under wind during its lifetime.

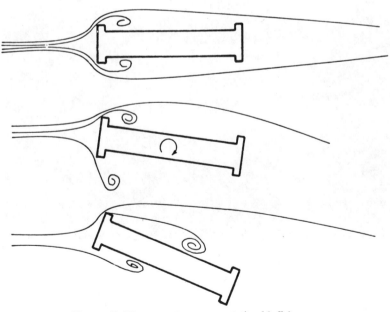

Figure 7. Vortex pattern over rotating bluff form.

FIGURE 8. Bridge section model in FHWA wind tunnel.

FIGURE 9.   Bosporus Bridge model in NPL wind tunnel, U.K.

Two of the most important flutter derivatives obtained from the section model test are H^*_1, a nondimensional coefficient proportional to the damping in vertical bending, and A^*_2, a similar coefficient proportional to the damping in torsion. Plots of the evolution of these coefficients as functions of V/NB, where N is the test oscillation frequency, serve to identify both the vortex-associated and flutter tendencies of the bridge deck section under small, incipient motions.

FIGURE 10 † typifies values of these coefficients as well as all aerodynamic coupling coefficients,[8] for the original Tacoma Narrows bridge, with airfoil

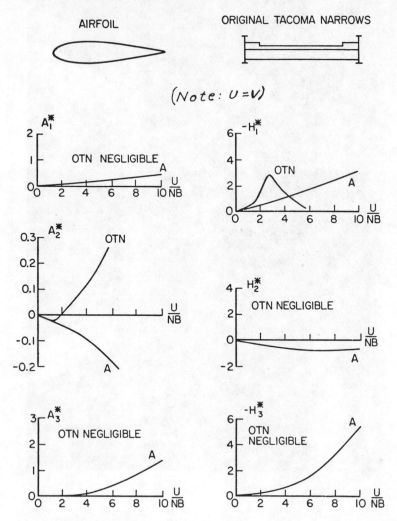

FIGURE 10. Flutter coefficients: original Tacoma Narrows and airfoil sections.

† In Figures 10–12 U and V are interchangeable.

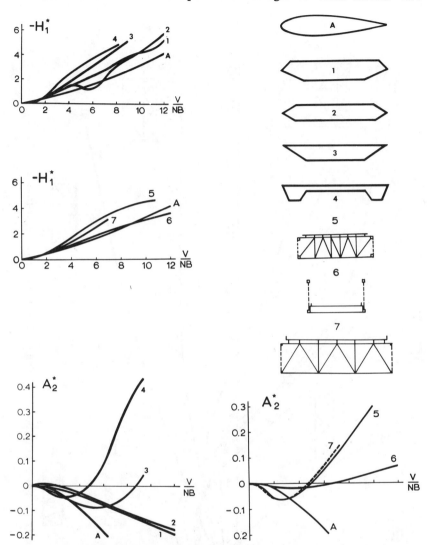

FIGURE 11. Vertical and torsional damping coefficients, airfoil, and 7 bridge forms.

results included for comparison. Note that coupling effects for the *OTN* are negligible, while for the airfoil they are strong. FIGURE 11 gives H^*_1 and A^*_2 for a set of 7 common bridge cross-sections. FIGURE 12 depicts a set of A^*_2 coefficients for a variety of stable and unstable truss-stiffened decks. The particularly striking aspect of the torsional damping coefficients A^*_2 is its tendency, with certain bridges, to reverse sign, from stable to unstable, for increasing values of V/NB, i.e. increasing wind velocity.

In spite of the fact that some bridges show such a reversal in A^*_2, they are not necessarily unstable *de facto* (a case in point being the Sitka Harbor

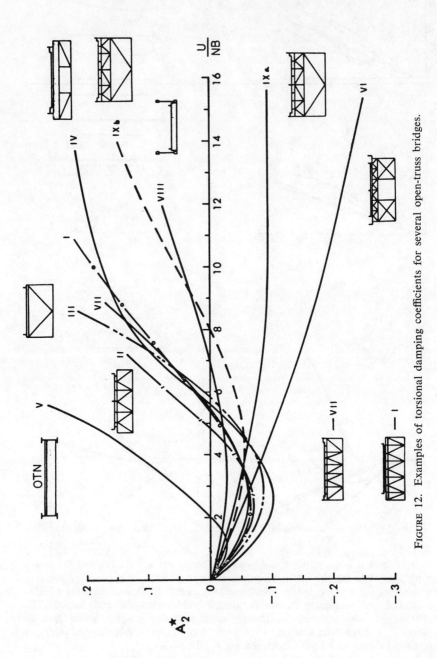

FIGURE 12. Examples of torsional damping coefficients for several open-truss bridges.

Bridge). This is accounted for simply by the fact that they may be torsionally stiff (i.e. N is a large number), so that V must also be high in order for flutter to occur. This emphasizes a point easily recognizable in bridge designs: a bridge may have an aerodynamically poor cross section but still be safe from flutter because of its torsional stiffness.

Thus high torsional stiffness clearly is an important flutter preventative. The Golden Gate Bridge is another case in point: its aerodynamic cross-section is quite flutter-prone, but it was torsionally stiffened to very good avail in the early 1950s, enhancing its flutter resistance. The Bronx-Whitestone similarly was torsionally stiffened by the addition of two side trusses during the early 1940s.

These measures, while effective, nevertheless may be classed with the "brute force" school of flutter prevention (that usually occurs as a *post facto* fix), whereas design studies concerned with proper geometric form adjustment prior to erection operate against the intrinsic mechanisms through which destabilizing aerodynamic forces are engendered. The latter is the preferable route for designers.

BUFFETING

The natural wind, passing over various obstacles of the local terrain, will generally be turbulent, and such turbulence buffets the bridge. Even if the bridge is designed so that it is flutter-free, it will still be subject to buffeting by gusts.

One way to assess the character of the bridge response to buffeting—and, indeed, to flutter and vortex-induced effects as well—is to build a scaled, full-bridge, aeroelastic model and test it in a wind tunnel under simulated turbulent wind. This has been accomplished on a few occasions and at considerable expense. A full-bridge wind-tunnel model of this sort, when placed in an environment that properly simulates natural wind turbulence, serves as a complete analog computer of the entire bridge performance under wind.

Approaches seeking to utilize to the fullest the data derivable from section models have sought to circumvent the cost in time and money of full models by calculating expected flutter and buffeting response via equations of motion employing the flutter derivatives. Such methods have been used in response studies of the Lions' Gate and Ruck-a-Chucky bridges by researchers at Princeton. Papers by Scanlan and Gade [14] and Scanlan [15] have discussed the underlying theory, incorporating the flutter derivatives. The state of both experimental and calculational art in regard to buffeting has reached a stage where good analytical assessments of buffeting response for a long-span bridge can be made. (cf. refs. 14–16). These can be useful in preliminary design.

The Golden Gate Bridge underwent serious buffeting in the earlier 1950s. The Deer Isle Bridge, though stiffened by various stay-cable configurations, continues to be active in wind, as does the Bronx-Whitestone. Broadly speaking it can be asserted that a bridge rendered secure against flutter by aerodynamic (as against structural) means may not be expected to undergo severe buffeting response in high winds because its potentially destabilizing torsional damping derivative (A^*_2) remains well-behaved. Contrarily, if the bridge is flutter-resistant merely through torsional structural stiffening, but is aerodynamically unsound, it may undergo severe buffeting response at high wind velocities

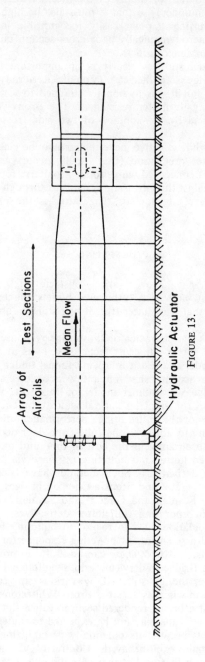

Test Sections

Mean Flow

Array of Airfoils

Hydraulic Actuator

FIGURE 13.

because its total damping is reduced toward zero aerodynamically, as wind velocity increases, negating its inherent structural damping.

Thus there remain very strong incentives for the designer of a long-span bridge to explore all aerodynamic avenues favorable to stability that can be associated with bridge-deck contour. There also is a growing trend to monitor

FIGURE 14.   CSU-Princeton 2D active gust generator.

new full-scale designs under wind action to assess their actual performance after erection.

A few final remarks on model testing and its relation to the buffeting problem are in order. Tests of section models have most often been made under laminar incident flow. This has proven to be conservative, as most realistic turbulent flows appear to delay to high windspeed the onset of the flutter observed under laminar conditions. Recently, however, reports have come in from at least two sources (H. Teunissen,[17] remarking on wind conditions at the Lions' Gate Bridge, Vancouver, and F. Durgin,[18] discussing flow conditions high over Boston) to the effect that natural wind may also on occasion exhibit unexpectedly laminar character. Under these conditions one cannot rely upon natural turbulence to delay or prevent flutter.

Thus, testing under laminar flow remains a conservative practice. A recent research study, however, jointly between Princeton University and Colorado State University ‡ is establishing an actively driven two-dimensional gust generator in a wind tunnel to be used for testing section bridge models. Figures 13 and 14 suggest aspects of this long-wavelength gust generator that is expected to produce suitably scaled turbulence that cannot otherwise be set up by purely passive means. It is planned to measure flutter derivatives and other characteristics of bridge deck sections with this facility. This constitutes the latest in a long series of research devices (cf. ref. 21) aimed at the proper solution of the potentially catastophic or annoying wind-induced responses of long-span bridges.

Conclusion

At the present time the theory and methodology for dealing adequately with the problems of vortex shedding, flutter, and the buffeting of bridges have been developed to a point of strong effectiveness. Wind tunnel testing, particularly of deck section models, can provide the designer of long-span bridges with sufficient information to influence designs in the direction of safety against wind-induced oscillations. With a minimum of careful basic testing, followed by sound analytical studies, designs can be evolved which are fundamentally secure against catastrophic wind-induced events and for which annoying or fatigue-related responses can be adequately predicted and minimized.

References

1. Yakubovich, V. A. & V. M. Starzhinskii. 1975. Linear Differential Equations with Periodic Coefficients. J. Wiley & Sons. New York. Chapter VI. (Transl. from Russian.)
2. Farquharson, F. B., Ed. Aerodynamic Stability of Suspension Bridges. University of Washington Engrg. Exp. Sta. Bull. No. 116, Parts I-V, June 1949–June 1954.
3. Scanlan, R. H. & R. L. Wardlaw. 1973. Reduction of flow-induced structural vibrations. In Isolation of Mechanical Vibration, Impact, and Noise. Am. Soc. Mech. Eng. pp. 35–63.
4. Gade, R. H. & H. R. Bosch. 1977. Aerodynamic Investigations of the Luling, Louisiana Cable-Stayed Bridge. Report FHWA–RD–77–161, Offices of Research and Development, Federal Highway Administration, U.S. DOT, Washington, DC, October.
5. Selberg, A. & E. Hjorth-Hansen. 1966. Aerodynamic stability and related aspects of suspension bridges. Proc. Symp. on Suspension Bridges. Lisbon, Nov.
6. Wardlaw, R. L. 1970. Static Force Measurements of Six Deck Sections for the

‡ Sponsored by the Federal Highway Administration of the DOT.

Proposed New Burrard Inlet Crossing. Rept. LTR–LA–53, NAE, National Research Council, Ottawa, Canada.
7. WARDLAW, R. L. 1974. A Wind Tunnel Study of the Aerodynamic Stability of the Proposed Pasco-Kennewick Intercity Bridge. Rept. LTR–LA–163, NAE, National Research Council, Ottawa, Canada, July.
8. SCANLAN, R. H. 1975. Recent Methods in the Application of Test Results to the Wind Design of Long, Suspended-Span Bridges. Report FHWR–RD–75–115, Offices of Research and Development, Federal Highway Administration, U.S. DOT, Washington, DC, October.
9. SCRUTON, C. 1948. Severn Bridge wind tunnel tests. Surveyor (London), **107**, No. 2959: 555.
10. DAVENPORT, A. G. et al. 1970. A Study of Wind Action on a Suspension Bridge During Erection and Completion. Rept. BLWT, 3–69 and 4–70, Fac. of Eng. Sci., University of Western Ontario, London, Canada, May 1969 and March 1970.
11. IRWIN, H. P. A. H. 1977. Wind Tunnel and Analytical Investigations of the Response of Lions' Gate Bridge to a Turbulent Wind. Rept. LTR–LA–210, NAE, National Research Council, Ottawa, Canada, June.
12. MELBOURNE, W. 1978. West Gate Bridge Wind Tunnel Tests. Internal Rept., Dept. of Mech. Eng. Monash University, Clayton, Victoria, Australia.
13. OLIVARI, D. & F. THIRY. 1975. Wind Tunnel Tests on the Aeroelastic Stability of the Heer-Agimont Bridge. Tech. Note 113, Von Karman Institute for Fluid Dynamics, Rhode-Saint-Genese, Belgium.
14. SCANLAN, R. H. & R. H. GADE. 1977. Motion of suspended bridge spans under gusty wind. J. Structure Division, Am. Soc. Civ. Eng. **103**, No. ST9:1867–1883.
15. SCANLAN, R. H. 1978. The action of flexible bridges under wind. I: Flutter theory; II: Buffeting theory. J. Sound and Vibration **60**, No. 2:187–199; and :201–211.
16. DAVENPORT, A. G. 1962. Buffeting of a suspension bridge by storm winds. J. Structural Division, Am. Soc. Civ. Eng. **88**, No. ST3:233–268.
17. TEUNISSEN, H. Private communication.
18. DURGIN, F. Private communication.
19. SIMIU, E. & R. H. SCANLAN. 1978. Wind Effects on Structures. John Wiley & Sons. New York.
20. WALSHE, D. E. (National Maritime Institute, U.K.). Private communication.
21. SCANLAN, R. H. 1979. On the state of stability considerations for suspended-span bridges under wind. Proc. IUTAM/IAHR Symposium on Flow-Induced Oscillations. University of Karlsruhe, Sept. In press.

AERODYNAMIC LESSONS LEARNED FROM INDIVIDUAL BRIDGE MEMBERS

CARL C. ULSTRUP

Only one Tacoma Narrows Bridge disaster was necessary to make engineers fully aware that they must deal with aerodynamics as well as aerostatics in the overall design of a bridge. In a less spectacular but also important segment of the design problem a lesson in aerodynamics has been taught on many occasions, but as shown by recent events, not fully learned.

I refer to the aerodynamic effect of wind action on individual bridge members.

Wind-induced vibration in individual bridge members follows a familiar pattern and some case histories will be mentioned.

The Askerödfjord Bridge in Sweden is an arch with cylindrical columns supporting the deck (PLATE 1). When it opened in 1960 the tallest columns were found to vibrate intensively at certain wind velocities. The remedy decided upon was to fill the cylinders with sand. Whether this proved to be successful has apparently not been published.

When a similar bridge was completed in the mid-1960s in Czechoslovakia it had the same problem (PLATE 2). An article published about this occurrence stated that filling the columns with sand was contemplated.

The Peace River Bridge in Canada consists of two tied arches with cylindrical hangers. Shortly after the first span was erected in 1967 the longest hangers were found to vibrate intensively at wind velocities as low as 10 miles per hour. Fatigue failure in the connections was a definite possibility.

The remedy of sand filling was first investigated. A model consisting of a 2 inch diameter pipe, 8 feet long was tested and it was found that when sand was added the oscillations died out more quickly.

The longest hanger in the bridge was then filled with sand. The result was just the opposite of what was expected. The critical wind velocity was lowered and the oscillations extended. Sand filling was not the solution. The increase in mass without increase in rigidity therefore lowered the frequency of vibration.

Installation of helical spoilers was contemplated, but that required welding on the tension member and was abandoned. The solution finally chosen was longitudinal cable ties between the hangers.

The Canadian experience raises the question whether the Swedish and Czech bridges needed further remedies to stabilize them.

The Tacony Palmyra Bridge over the Delaware River is a tied-arch bridge that was opened to traffic in 1929. The hangers consisted of H-shaped sections. Vibration problems occurred immediately. The remedy was to install one line of longitudinal bracing.

In the late 1930s some of the H-shaped hangers on the tied-arch Storeström

Carl C. Ulstrup is an Associate With Steinman, Boynton, Gronquist & Birdsall, 50 Broad Street, New York, New York 10004.

0077–8923/80/0352–0265 $01.75/1 © 1980, NYAS

PLATE 1. Askerödfjord Bridge, Sweden.

PLATE 2. Orlik Bridge, Czechoslovakia.

Bridge in Denmark fractured in the connections during construction. The H sections were replaced by a set of battened channel sections.

The Fire Island Bridge on Long Island is a tied arch with H-shaped hangers. When the bridge opened in 1964 the hangers vibrated. Bent plates were riveted to the entire length of the flanges, and apparently worked as spoilers or stiffeners.

A tied-arch type bridge with diagonal cylindrical hangers was completed in Japan in 1967. The longer hangers were found to vibrate violently, and

PLATE 3. Commodore John J. Barry Bridge over Delaware River.

cracks developed at the connections. The counter measures were to reinforce the connections to create near fixity and in addition wires were wound in a helical fashion around the members to act as spoilers.

The longest cantilever bridge in the United States is the Commodore John J. Barry Bridge over the Delaware River (PLATE 3). Its main span is 1,644 feet. During construction in 1973 the flanges of nine H-shaped members were found to be almost completely cracked at the connections after a two-day

PLATE 4. Cracked hanger after temporary repairs.

wind storm. The immediate remedy was to splice the flanges and to install cable ties (PLATES 4–8). Both the owner and the contractor engaged consulting engineers to determine the cause of the cracking. Wind tunnel tests were performed on sectional models and aeroelastic models geometrically scaled. They both verified the computed critical velocities.

PLATE 5. Another cracked hanger after temporary repairs.

PLATE 6. Temporary longitudinal and transverse wire rope bracing.

One of the consultants, Steinman, Boynton, Gronquist & Birdsall, selected a member for testing whose actual dimensions were as follows: Length 140 feet, flanges ⅝ × 36 inches and web ⁷⁄₁₆ × 27 inches. Assuming the ends of this tension member to be fixed, the computed critical wind velocity was found to be 44 miles per hour. This was a wind velocity that frequently had to be expected at the bridge site.

An aeroelastic model with a geometric scale of 8.36 was tested in the Boeing Vertol wind tunnel (PLATE 9). At about 40 miles per hour wind velocity the model began to vibrate so violently that the test had to be stopped to prevent the model's destruction. Both weak axis and torsional vibration were observed. In another test on a sectional model in a wind tunnel at Virginia Polytechnic Institute resonant vibration occurred at about 40 miles per hour both for weak axis bending and torsion (PLATE 10).

The final remedy decided upon was to install vibrational dampers on 258 of the bridge members at an approximate cost of 1.3 million dollars. The dampers had proved effective when installed on the Bras d'Or Bridge in Canada in 1974. This bridge had I-beam hangers that suffered damage after the bridge was completed.

So from country to country, time and again the annoying problem of wind-induced vibration pops up. Costly, sometimes questionable and often unsightly remedies must be undertaken, usually followed by lawsuits.

Why do bridge designers often have little success in checking the critical wind velocity of hangers and columns, the failure of which may be catastrophic? The answer may be best explained as follows:

Lengthy reports on wind tunnel tests and mathematical theories which attempt to explain this phenomenon have been published. The problem is they seem to be mere exchanges between researchers rather than information for the practicing engineer.

With nothing better to guide him than a paper reporting on the investigation of a bridge with vibration problems, the designer may find himself in a mathematical jungle of formulas involving flutter, buffeting, galloping, and vortex shedding. Attempts made to compute vibration amplitudes and resulting stresses based on such literature could be misleading.

The three most common shapes prone to vibration are H sections, box sections, and cylinders. The latter is known as a stable section since it will vibrate with a finite amplitude at critical wind velocities. But fatigue failure in the connections is a definite possibility. The first two sections are unstable and may experience destructive oscillations.

PLATE 7. Temporary bracing.

PLATE 8. Temporary bracing.

It is significant that in all the case histories mentioned the simple formula for vortex shedding will accurately predict the critical wind velocities that will cause the member to vibrate.

Vortex-induced oscillation is the most common form of wind-induced vibration in engineering. In the presence of wind, vortices of equal strength but opposite rotation are shed from one side of the body and then from the other causing the member to vibrate in a plane normal to the wind.

The frequency of vortex shedding increases with the wind velocity. When it coincides with the natural frequency of the member at the so-called critical wind velocity, resonance occurs and violent vibration may result. To the design engineer it is really not important to know the intensity of vibration. What he wants to know is that no resonant vibration should take place at wind velocities under 100 miles per hour.

In order to compute the critical wind velocity two things have to be known about the member: namely, its Strouhal Number and its natural frequencies. The value of the Strouhal Number has been determined experimentally for a great number of sections and can be found in readily available publications.[1] The natural frequency of the member depends on the section properties and the end-support conditions. The presence of a tensile force will cause an increase in the frequency, whereas a compressive force will cause a decrease. There are a large number of natural frequencies for a member, analogous to buckling modes for a column. It may vibrate in the first, second, third, etc. mode.

The natural frequencies of a member subject to axial forces can in certain cases be determined by what might be called exact formulas. In most cases these formulas involve iteration and are therefore properly subject for solution by computer. However, some quite simple short cut formulas exist, which enable the engineer to quickly determine the frequencies on a calculator.

In the published appendix to this paper such formulas appear. They assume the member will vibrate in the buckled shape of a corresponding column and apply to both bending and torsional vibration and to any support condition at the ends. The buckling analogy enables the engineer to visualize the deflected shape of the member. If the critical wind velocity is found to be too low for safe design, the member can be forced into one of the higher modes of vibration, with a higher critical wind velocity, by installing bracing at the nodal points

PLATE 9. A 1:8.36 scale hanger model being tested in Boeing Vertol's wind tunnel.

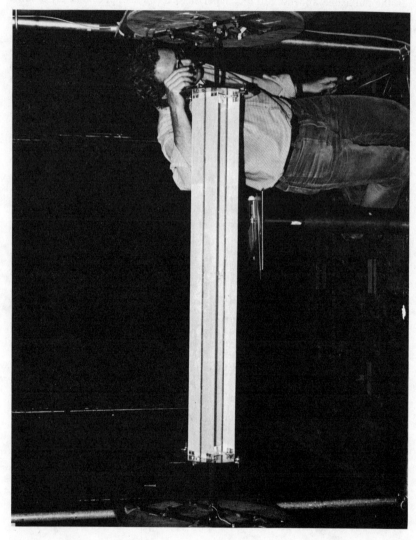

PLATE 10. Sectional model of a hanger being tested in a wind tunnel at Virginia Polytechnic Institute.

or changing shape or stiffness. A numerical example illustrates the use of the formulas.

Although not the point of this paper, having ascertained that all is well vibration-wise, the designer should take a look at one more thing. He should check the static stresses caused by the wind load. They may turn out to be considerable.

APPENDIX

The frequency of vortex shedding is determined by the formula

$$f = s\frac{V}{d} \tag{1}$$

where S = Strouhal number
V = Wind velocity in feet per second or meters per second
d = Characteristic width of cross section measured normal to the direction of the wind

The value of the Strouhal number has been determined for many sections some of which are shown in TABLE 1.

The natural frequencies of a member subject to an axial force can be determined closely by the following formula (see reference 5).

$$f_n = \frac{1}{2\pi}\left(\frac{k_n\ell}{\ell}\right)^2 a\left[1 + e_p\left(\frac{K\ell}{\pi}\right)^2\right]^{\frac{1}{2}} \tag{2}$$

where f_n = natural frequency of member for each mode corresponding to $n = 1, 2, 3$ etc.
$k_n\ell$ = eigenvalue for each mode and is listed in TABLE 2
K = effective length factor and is also listed in TABLE 2
ℓ = length of the member
a = coefficient dependent on the physical properties of the member
e_p = coefficient dependent on the physical properties of the member and on the axial force

This formula applies to both bending and torsional vibration and to any support conditions at the ends. It is limited to members whose shear center and centroid coincide.

For bending
$$a_b = \left(\frac{EIg}{\gamma A}\right)^{\frac{1}{2}} \tag{3}$$

$$e_{pb} = \frac{P}{EI} \tag{4}$$

For Torsion
$$a_t = \left(\frac{EC_wg}{\gamma I_p}\right)^{\frac{1}{2}} \tag{5}$$

$$e_{pt} = \frac{GJ + PI_pA^{-1}}{EC_w} \tag{6}$$

TABLE 1

STROUHAL NUMBER FOR VARIOUS SECTIONS
(From reference 1, except as noted.)

Wind	Profile Proportion	Value of S	Profile Proportion	Value of S
→	d, d	.120	(circle)	.200
↓		.137		
↓	$\frac{1}{2}$d, d	.144	b, d	
↓	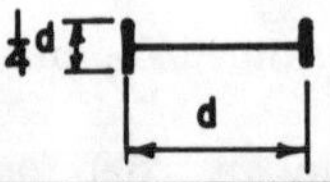 $\frac{1}{4}$d, d	.145	From reference (2) b/d 2.5	.060
			2.0	.080
			1.5	.108
↓	$\frac{1}{4}$d, $\frac{1}{2}$d, $\frac{1}{4}$d, d	.147	1.0	.133
			.7	.136
			.5	.138

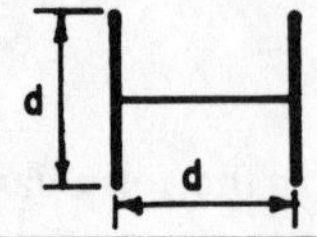

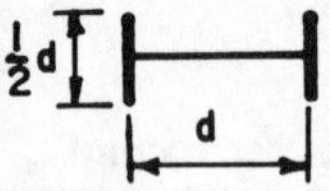

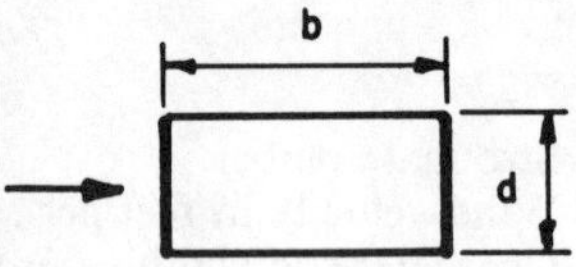

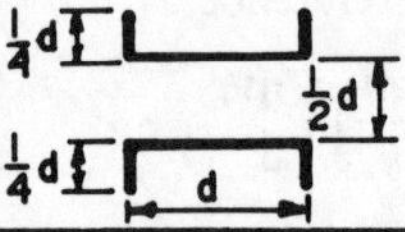

TABLE 2

EIGENVALUE $k_n\ell$ AND EFFECTIVE LENGTH FACTOR K
(Values of $k_n\ell$ from references 3 and 4.)

Support Condition	$k_n\ell$			K		
	n=1	n=2	n=3	n=1	n=2	n=3
 (pin-pin)	π	2π	3π	1	$\frac{1}{2}$	$\frac{1}{3}$
 (fixed-pin)	3.927	7.069	10.210	.7	.412	.292
 (fixed-fixed)	4.730	7.853	10.996	.5	.35	.259
 (fixed-free)	1.875	4.694	7.855	2	$\frac{2}{3}$	$\frac{2}{5}$

Special Case: $C_w = O$

$$f_n = \frac{1}{2\pi}\left(\frac{k_n\,\ell}{\ell}\right)^2\left[\frac{g}{\gamma I_p}(GJ + PI_p\,A^{-1})\left(\frac{K\,\ell}{\pi}\right)^2\right]^{1/2} \tag{7}$$

where E = Young's modulus
$\quad G$ = shear modulus
$\quad \gamma$ = weight density of member
$\quad g$ = gravity acceleration
$\quad P$ = axial force (tension is positive)
$\quad I$ = moment of inertia about relevant axis
$\quad A$ = area of member cross section
$\quad C_w$ = warping constant
$\quad J$ = torsion constant
$\quad I_p$ = polar moment of inertia

NUMERICAL EXAMPLE—TENSION MEMBER

Section properties for member shown in FIGURE 1:

$\quad A = 58.97$ in^2 (380 cm^2)

$\quad I_x = 10,085$ in^4 (419,770 cm^4) $\quad I_y = 4,860$ in^4 (202,290 cm^4)

$\quad I_p = I_x + I_y = 14,945$ in^4 (622,060 cm^4)

$\quad J = \frac{1}{3}\sum st^3 = \frac{1}{3}\left[2 \times 36 \times \left(\frac{5}{8}\right)^3 + 28\frac{9}{16}\left(\frac{1}{2}\right)^3\right] = 7.05$ in^4 (293 cm^4)

$\quad C_w = \frac{1}{4}I_y h^2 = \frac{1}{4} \times 4,860\left(28\frac{9}{16}\right)^2 = 991,217$ in^6 (2.66 $\times$ 10^8 cm^6)

$\quad S_x = 691$ in^3 (11,323 cm^3) $\quad S_y = 270$ in^3 (4,425 cm^3)

$\quad r_y = 9.08$ in (23 cm)

$\quad g = 32.16$ ft/sec^2 (9.80 m/sec^2)

Material Properties:

$\quad E = 29 \times 10^6$ psi (2 $\times$ 10^8 kN/m^2) $\quad G = \dfrac{E}{2.6}$ $\quad \gamma = 490$ pcf (7,850 kg/m^3)

Given:

$\quad \ell = 140.5$ ft $= 1,686$ in (42.82 m) $\quad P = 627,000$ lbs. (2,790 kN)

$\quad \dfrac{\ell}{r_y} = \dfrac{1,686}{9.08} = 186 < 200$

Compute weak axis bending frequencies as well as torsional frequencies.

$\quad a_b = \left(\dfrac{EI_y g}{\gamma A}\right)^{1/2} = \left(\dfrac{29 \times 10^6 \times 4,860 \times 32.16}{490 \times 58.97}\right)^{1/2}$
$\quad\quad = 12,525$ ft^2/sec (1,164 m^2/sec)

$\quad e_{pb} = \left(\dfrac{P}{EI_y}\right) = \dfrac{627,000}{29 \times 10^6 \times 4,860} = 4.4487 \times 10^{-6}$ in^{-2} (6.893 $\times$ 10^{-3} m^{-2})

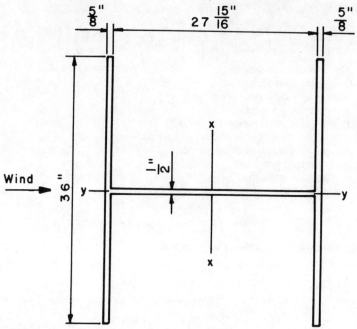

FIGURE 1

$$a_t = \left(\frac{EC_w g}{\gamma I_p}\right)^{1/2} = \left(\frac{29 \times 10^6 \times 991,217 \times 32.16}{490 \times 14,945}\right)^{1/2}$$
$$= 11,236 \text{ ft}^2/\text{sec } (1,044 \text{ m}^2/\text{sec})$$

$$e_{pt} = \frac{GJ + PI_p A^{-1}}{EC_w} = \frac{29 \times 10^6 \times 2.6^{-1} \times 7.05 + 627,000 \times 14,945 \times 58.97^{-1}}{29 \times 10^6 \times 991,217}$$
$$= 8.263 \ 10^{-6} \text{ in}^{-2}$$
$$(.0128 \text{ m}^{-2})$$

$$f_n = \frac{1}{2\pi}\left(\frac{k_n \ell}{\ell}\right)^2 a\left[1 + e_p\left(\frac{K\ell}{\pi}\right)^2\right]^{1/2}$$

Consider member with fixed ends and compute the two lowest frequencies.

Bending: $f_{n=1} = \frac{1}{2\pi}\left(\frac{4.730}{140.5}\right)^2 \times 12,525\left[1 + 4.4487 \times 10^{-6}\left(\frac{.5 \times 1,686}{\pi}\right)^2\right]^{1/2}$

   $f_{n=1} = 2.26 \times 1.1491 = 2.60 \text{ hertz}$

   $f_{n=2} = \frac{1}{2\pi}\left(\frac{7.853}{140.5}\right)^2 \times 12,525\left[1 + 4.4487 \times 10^{-6}\left(\frac{.35 \times 1,686}{\pi}\right)^2\right]^{1/2}$

   $f_{n=2} = 6.23 \times 1.0756 = 6.70 \text{ hertz}$

Torsion:
$$f_{n=1} = \frac{1}{2\pi}\left(\frac{4.730}{140.5}\right)^2 \times 11{,}236\left[1 + 8.2635 \times 10^{-6}\left(\frac{.5 \times 1{,}686}{\pi}\right)^2\right]^{1/2}$$

$$f_{n=1} = 2.03 \times 1.2629 = 2.56 \text{ hertz}$$

$$f_{n=2} = \frac{1}{2\pi}\left(\frac{7.853}{140.5}\right)^2 \times 11{,}236\left[1 + 8.2635 \times 10^{-6}\left(\frac{.35 \times 1{,}686}{\pi}\right)^2\right]^{1/2}$$

$$f_{n=2} = 5.59 \times 1.1365 = 6.35 \text{ hertz}$$

TABLE 3 shows the three lowest frequencies obtained from the approximate general formula for weak axis and torsional vibration of the same member with four different end conditions. The values in brackets were obtained by computer from the "exact" formulas.

Vibration and bending about the weak axis is caused by the wind blowing on the flange. However, static bending about the strong axis also is produced by this wind force. The combined static axial and bending stresses will be maximum at the edge of one flange.

Determine the static bending stress f_b for a wind load of 50 #/ft² (2,395 N/m²) (AASHTO Specifications) and the axial stress f_a for $P = 627{,}000$ lbs. (2,790 kN).

$$q = 50 \times 3 = 150 \text{ #/ft (2,190 N/m)} \quad U^2 = \frac{P\ell^2}{4EI_x} =$$

$$\frac{627 \times 1{,}686^2}{4 \times 29 \times 10^3 \times 10{,}085} = 1.5235$$

Member with Hinged Ends

$$y_{\mathbb{C}} = \frac{5}{384}\frac{q\ell^4}{EI}\frac{\frac{1}{CoshU}-1+\frac{1}{2}U^2}{\frac{5}{24}U^4} = 4.497 \times .617 = 2.77 \text{ in (70.4 mm)}$$

TABLE 3

NATURAL FREQUENCES FOR VARIOUS END CONDITIONS

	Natural Frequencies in Hertz			
	Hinged Supports	Fixed—Fixed	Fixed—Hinged	Cantilever
(1)	(2)	(3)	(4)	(5)
Bending				
$f_{n=1}$	1.51 [1.51]	2.60 [2.58]	1.99 [1.97]	0.88
$f_{n=2}$	4.58 [4.58]	6.70 [6.69]	5.57 [5.57]	2.79
$f_{n=3}$	9.59 [9.59]	12.72 [12.72]	11.09 [11.09]	6.84
Torsion				
$f_{n=1}$	1.64 [1.64]	2.56 [2.54]	2.06 [2.03]	1.03
$f_{n=2}$	4.52 [4.52]	6.35 [6.33]	5.36 [5.36]	2.86
$f_{n=3}$	9.05 [9.05]	11.80 [11.79]	10.36 [10.36]	6.57

$$M_{\mathbb{C}} = \frac{1}{8} q \ell^2 \frac{2\,(CoshU-1)}{U^2 CoshU} = 370{,}130 \times .6083 = 225{,}151 \text{ ft-lbs } (305.4 \text{ kN.m})$$

$$f_b = \frac{225{,}151 \times 12}{691} = 3{,}910 \text{ psi} \qquad\qquad f_a = \frac{627{,}000}{58.97} = 10{,}633 \text{ psi}$$
$$(26{,}980 \text{ kN/m}^2) \qquad\qquad\qquad\qquad\qquad (73{,}368 \text{ kN/m}^2)$$

Maximum static flange stress = 14,543 psi at midspan
(100,348 kN/m²)

Additional stress may be caused by vibration.

Member with Fixed Ends

$$y_{\mathbb{C}} = \frac{1}{384} \frac{q\,\ell^4}{EI} \frac{24}{U^4} \left(\frac{1}{2} U^2 - \frac{U\,CoshU - U}{SinhU} \right) = .899 \times .868 = .78 \text{ in } (19.8 \text{ mm})$$

$$M_{sup} = -\frac{1}{12} q\,\ell^2 \frac{U\text{-}TanhU}{\frac{1}{3} U^2 TanhU} = -246{,}753 \times .9112 = -224{,}847 \text{ ft-lbs}$$
$$(-305.0 \text{ kN.m})$$

$$f_b = \frac{224{,}847 \times 12}{691} = 3{,}905 \text{ psi} \qquad\qquad f_a = 10{,}633 \text{ psi}$$
$$(26{,}945 \text{ kN/m}^2) \qquad\qquad\qquad\qquad (73{,}368 \text{ kN/m}^2)$$

Maximum static flange stress = 14,538 psi (100,313 kN/m²)

For a different value of tension in the member the moment $M_{\mathbb{C}}$ for the hinged end condition would not be the same as M_{sup} for the fixed end condition.

Together with the static stresses, additional stresses are produced by vibration. The frequency formulas do not yield the amplitude y of oscillation, but it is possible by means of wind tunnel tests to predict the actual deflection of the vibrating member. A stable member will have a finite deflection from which the stresses can be computed. On the other hand, an unstable member may either vibrate with a finite amplitude or it may vibrate so violently that it will fail. The determination of the amplitude and stress produced by vibration is not included in this paper.

Critical Wind Velocities

When the frequency of vortex shedding equals one of the natural frequencies of the member, resonance occurs and the corresponding wind velocity is the critical wind velocity = V_{cr}

$$V_{cr} = \frac{f_n d}{S} \qquad\qquad (1)$$

For weak axis bending and torsion use $d = 36'' = 3'$ (91.4 cm) and the Strouhal Number $S = .12$.

The member with hinged ends has $f_{n=1} = 1.51$ hertz

$$V_{cr} = \frac{1.51 \times 3}{.12} \times \frac{3{,}600}{5{,}280} = 25.7 \text{ mph } (41.4 \text{ km/hr})$$

TABLE 4 shows the critical wind velocities for weak axis and torsional vibration for members with four different support conditions.

For each of these support conditions the frequencies are shown for three modes of vibrations: without nodal points ($n = 1$), with one nodal point ($n = 2$) and with two nodal points ($n = 3$).

All members except the one with fixed ends should be braced at the third points to be safe in a 100-mph wind. The fixed end member should be braced at midpoint, corresponding to $n = 2$.

TABLE 4

CRITICAL WIND VELOCITIES FOR VARIOUS END CONDITIONS

| (1) | Critical Wind Velocities in MPH * | | | |
	Hinged Supports	Fixed—Fixed	Fixed—Hinged	Cantilever
(1)	(2)	(3)	(4)	(5)
Bending				
$f_{n=1}$	25.7	44.3	33.9	15.0
$f_{n=2}$	78.1	114.2	94.9	47.6
$f_{n=3}$	163.5	216.8	189.0	116.6
Torsion				
$f_{n=1}$	28.0	43.6	35.1	17.6
$f_{n=2}$	77.0	108.2	91.4	48.8
$f_{n=3}$	154.3	201.1	176.6	112.0

* 1 mph = 1.61 km/hr.

REFERENCES

1. TASK COMMITTEE ON WIND FORCES. 1961. Wind Forces on Structures. Trans. Am. Soc. Civ. Eng., **126**, Part II.
2. PARKINSON, G. V. 1965. Aeroelastic Galloping in One Degree of Freedom, Wind Effects on Buildings and Structures. National Physical Laboratory, Teddington, England. H. M. Stationery Office, London.
3. JACOBSEN, L. S. & R. S. AYRE, EDS. 1958. Engineering Vibration, with Application to Structures and Machinery. McGraw-Hill Book Company. New York.
4. TIMOSHENKO, S. & D. H. YOUNG, EDS. 1955. Vibration Problems in Engineering. 3rd edit. D. Van Nostrand Company, Inc. New York.
5. ULSTRUP, C. C. 1978. Natural Frequencies of Axially Loaded Bridge Members. J. Structural Division, Proc. Am. Soc. Civ. Eng., **104**, ST2.